I0636610

C. G Elliott

Practical Farm Drainage

Why, When and How to Tile Drain

C. G Elliott

Practical Farm Drainage
Why, When and How to Tile Drain

ISBN/EAN: 9783744678469

Printed in Europe, USA, Canada, Australia, Japan

Cover: Foto ©berggeist007 / pixelio.de

More available books at **www.hansebooks.com**

FARM DRAINAGE:

WHY, WHEN AND HOW TO TILE DRAIN.

BY

C. G. ELLIOTT,

ILLUSTRATED.

INDIANAPOLIS, IND.:

J. J. W. BILLINGSLEY, PUBLISHER.

1882.

PREFACE.

THE following pages are presented to those who are interested in the subject of farm drainage. So short has been the time since the introduction of tile drainage in the prairie States, that farmers have scarcely taken time to acquire the requisite knowledge of the subject before they have begun actual operations. The drainage practice of the eastern States must be adapted to western soil and surroundings. This requires time and close observation. The object of these few pages is to give, in a concise and plain manner, that which the farmer should know, if he contemplates draining his farm. It is not the intention of the author to say all there is to be said upon the subject, but to say enough to give the farmer an elementary knowledge of why, when, where and how to drain his farm. The practical methods described in these pages have been well tested, and are now in constant use by practical men. It is hoped that the language is sufficiently clear to be understood by all.

TONICA, ILLINOIS,   C. G. ELLIOTT,
    September, 1882.   *Civil Engineer.*

CONTENTS.

CONTENTS.

CHAPTER VI.

DITCHING MACHINES.

CHAPTER VII.

COST AND PROFIT.

CHAPTER VIII.

ROAD DRAINAGE.

LIST OF ILLUSTRATIONS.

CHAPTER I.

Introduction—Kinds of Land Requiring Drainage—Sources of Water—Mechanical Difference Between a Wet and a Dry Soil—The Relation of the Contour of the Surface and Sub-Soil to Drainage—Kinds of Drains—Open Drains—Tile Drains.

INTRODUCTION.

But very little attention has been given to land drainage in Illinois and other western states, until recently. A casual glance at our farms in the spring of the year, when many of them are partially submerged, and the farmer, with idle men and teams, is impatiently waiting for the slow natural drainage of flat land, and the evaporation of the rainfall by heat from the sun, before he can begin operations, will convince any observing man that the rapid removal of this surplus water would be of immense benefit to the agricultural community.

The practical feasibility of this work is at present the problem with many. The farmer asks himself and others, "Can I drain my field or my farm thoroughly, and will the probable returns justify the outlay?" Valid and useful conclusions upon this matter can not be arrived at until we have availed ourselves of the experience of others, and have obtained correct ideas of the principles of drainage—what thorough drainage is, and what it will accomplish.

It may be well to mention a few of the benefits accruing from drainage which are of actual money value to the farmer. These benefits are not hidden away in the soil, but may be seen by any one who will compare a well-drained field with one which is wet and un-drained.

First, there is no failure of crops on account of ex-cessive rains. Almost every farmer may put down among his losses the partial or total failure of several acres of land to produce a crop because, during some part of the season, the land was too wet.

Second, the soil is in condition to receive the crop at the proper season of the year, and it begins a healthy growth at once. This will add many dollars to the value of the field each year, and cost no more labor.

Third, the labor which produces a poor crop on un-drained land, will produce an excellent one on the same land when properly drained. In this way crops are often doubled on what is called average farm land.

Fourth, by reason of the absence of surplus water in the soil, grain and grass are not "heaved" and frozen out in winter time.

Fifth, whatever fertilizing material is put on the land is made more available for plant food, for the reason that the soil is more porous and not surface-washed, and fertilizers are at once incorporated in the soil. Undecayed matter put upon the soil decays more rap-idly and becomes sooner prepared for the use of plants. Fertilizing gases held in the air are carried by the rain into the soil, making it more rich, instead of being washed away or taken with vapor into the air again.

Other advantages will be mentioned as we proceed farther, but these just named will perhaps be sufficient to show the importance of the subject. Each season as it comes, turns another leaf of the book of Farm Econ-omy, telling the same story in different ways, and emphasizing it at times in such a manner as 'to compel the farmer to heed its teachings.

KINDS OF LAND REQUIRING DRAINAGE.

Ponds and sloughs are wholly unfit for cultivation, even in the dryest years, without drainage. *Ponds* are basins which seem to have been provided by nature for receptacles of surplus water flowing from the surrounding high land, and are reservoirs for the drainage of land which gives the farmer his profits. These ponds are generally covered with aquatic plants, which are very tenacious of life in wet soil, but easily killed when deprived of their natural nutriment by drainage. *Sloughs* are the natural water-courses of prairie land, being to level land what creeks and rivers are to more rolling and hilly sections. These are often broad and flat, allowing water to spread over rods of valuable land, where by suitable ditches it might be confined to a much narrower space and many acres of valuable land reclaimed.

Flat land under cultivation is usually the first land which directs the farmer's attention to draining. A season which is drier than usual shows to him that such soil, when not too wet, will produce a crop equal to his best fields. On this land the natural drainage is not rapid enough in the spring-time to fit it for the growth of plants. It is generally cultivated when too wet, which causes the soil to become compact and in time of drought it shrinks and cracks, resulting in the ruin of the crop and more than loss of the labor; for the soil is in a worse condition than it was in the spring before the plow was started.

Channels or runs through cultivated land often are common where the land is rolling. Water flows down the slopes and oozes from the banks until these runs are so wet that they rarely produce a crop, and are a great inconvenience in cultivating.

SOURCES OF WATER.

Primarily the source of all water of use or injury to the agriculturist is the rain-fall. Considered, however, with reference to drainage, we speak of *surface-water*, which rests upon the surface of the soil, a part passing down to the sub-soil, a part flowing over the surface and passing off, and the remainder raised by evaporation or used by plants; *ooze water*, which passes through the soil below its surface and finally rests in some channel or flat land, saturating it until it is unfit for cultivation; *spring water*, which has its source in some one locality of the field, or proceeds from some distant source through its own channels in the sub-soil. These must be provided for by drainage, according to the nature of the case.

MECHANICAL DIFFERENCE BETWEEN A WET AND A DRY SOIL.

If we look at a lump of dry soil by means of a common magnifying glass we see that it is made up of small particles thrown together miscellaneously, having small cavities between them resembling those of a sponge. The particles also have minute pores and cells which hold liquids by the power of absorption. The figures here given are drawn to illustrate this, and are similar in idea to those of Col. Waring, which he borrows from an English Report on Drainage. Let us dry a portion of soil and from it cut a small block. This, placed under a magnifying glass, will appear somewhat as represented in figure 1. It is composed of irregularly shaped particles having channels and cavities between them similar to those existing in a pile of small stones. These particles in turn have very minute cells, capable of absorbing and holding moisture. In the piece before us there is no moisture between the particles nor in them, both being filled with

Fig. 1.—A dry Soil

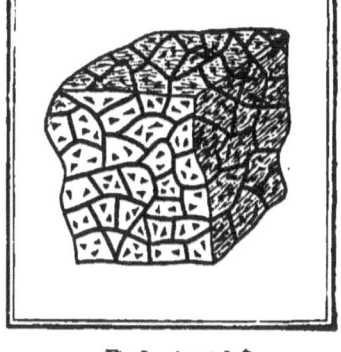

Fig. 2.—A wet Soil

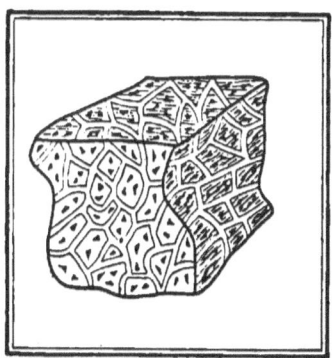

Fig. 3.—A drained Soil

air alone. It is evident from the laws of vegetable growth that such soil is unfit for the growth of seed or plant. If we pour water upon this block of dry soil until it will hold no more, we have the state of things shown in figure 2. The cavities, pores, cells, in short, every space before occupied by air is now filled with water. Seeds and roots in such a soil can not thrive, for all air is excluded, except what little may pass through the water to the growing roots.

If we notice the soil as we put the water upon the block under the glass, we will see that a drop placed upon one side or the top, changes the color of the soil, showing that the soil is moist, but there is no change in the spaces represented in the figure. This state is shown in fig. 3. The minute spaces in the particles are filled with moisture and will hold a certain percentage, varying with the kind of soil, while the spaces between the particles are filled with air. We have here an example of a drained soil in which the plant roots have access to both air and moisture. It will be seen, then, that draining is simply removing the surplus water from the soil. This allows the atmosphere to take its place, thereby giving the plant needed oxygen at its roots, and producing a chemical change in the soil which gives the plants more nourishment.

The amount of water held by absorption varies greatly with the kind of soil. To show that a well drained soil is by no means a dry soil, we have experiments by Professor Schübler, who found that one hundred pounds of dry soil would retain the following weight of water that would not flow off by drainage:

Sand ....25 pounds.
Loamy Soil.. ......40 pounds.
Clay Loam...50 pounds.
Pure Clay..70 pounds.

THE RELATION OF THE CONTOUR OF THE SURFACE AND SUB-SOIL
TO DRAINAGE.

A cross section of prairie soil usually shows, first, what is known as the soil, which consists of loam, more or less vegetable, to a depth of from 18 to 30 inches, then a few inches of mixed soil and clay, then fine clay,

varying in color according to locality. This consti-
tutes the sub-soil, and in its natural state is not so
easily permeated by water.

Natural drainage in such a soil is accomplished by
the surface water flowing down the slope, and the ooze
water flowing through the soil, most of it passing down
to the clay sub-soil, thence gradually oozes down the
slope until it finds some place of exit or level. By
reference to fig. 4 it will be seen that the relation of
surface and sub-soil has much to do with the facility
with which artificial drainage is effected.

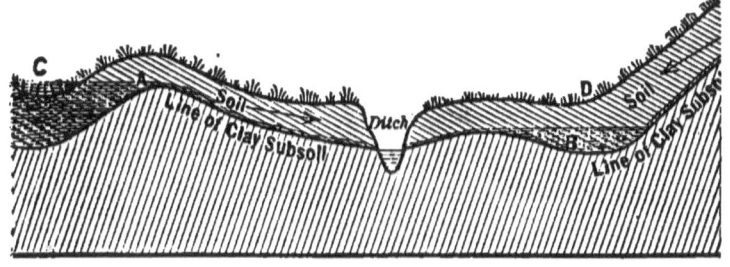

Fig. 4

At *A* the layer of clay rises nearer the surface than
at other places, holding the water back of it. until its
level becomes high enough to flow over the obstruction,
when it oozes down the regular slope and is discharged;
or as at *B*, the water after passing through the soil is
held below by the clay basin in quantities sufficient to
make the soil at *D* unfit for profitable cultivation.
Such is the case at *C*. We do not wish it understood
that sub-soil clay is impervious to water. It is only
comparatively so. . The soil is so susceptible to the
passage of water that the difference is all the more
marked. As seen before, the retentive power of soils
varies greatly. Understanding these natural difficulties, we can arrange our drainage with reference to

overcoming them. These things are mentioned, not with the intention of covering all cases, but to suggest to the thinking and investigating agriculturist something of what he should take into consideration when he undertakes to drain wet land.

KINDS OF DRAINS.

Many experiments have been made to find some inexpensive material for, and method of constructing, drains. None have stood the test of time but open ditches for surface drains, and tile pipes for under drains. If the western farmer wishes permanent and effective drainage, he must be at the expense of constructing suitable open ditches for large water-courses, and well laid lines of tile drains for general draining purposes. Drains constructed of boards, brush, gravel, etc., are less effectual and in the long run more expensive.

OPEN DRAINS.

However much open ditches may be disliked, they are often a necessity. The farmer who has experienced the convenience and profit of under-drains conceives the idea of doing away with all open ditches by using tiles and covering them, thus saving all inconvenience occasioned by the ditch, and also adding to his tillable land that occupied as a water-course. This operation will often retard the action of drains which discharge into the large channel, and if the slough is large will wholly prevent good drainage. As noticed before, sloughs on the prairies are the natural water-courses, give surface drainage to large tracts of land either side of them, and during seasons of heavy rain require large capacity in order to remove the water coming to them. In many cases a pipe eleven or twelve inches in diameter would do the work, but these are at present exceedingly expensive and can not often be used by individual farmers because of the expense.

Again, sloughs flowing through tracts of land which are flat and require under-draining more than any other, themselves have but little grade, so that very large pipes would be required to give the necessary discharge.

If all our land were well under-drained, thus preventing any accumulation of water in flat and low places in the bordering fields, or surface flooding, the case under consideration would be radically changed. The water, instead of requiring immediate removal as fast as it gathers upon the surface, would be taken up by a soil well drained to a depth of three or four feet and carried off by the pipes laid for that purpose. This drained soil is capable of holding a large quantity of water before the surface is covered, and the drains carry it off gradually. It will be seen that the main drain will not be called upon to carry as great a quantity of water as it would in case it were not supplemented by under-draining. In the opinion of the writer, open ditches must be used in all large sloughs where draining is no more thorough than is usually practiced, except about one mile at the head of the slough.

It must be borne in mind that the fall of the slough, the area drained by it, and the nature of the soil, greatly affect all our calculations and consequent practice in draining. The statements above made are perhaps as definite as can be given, and apply to all cases.

In the drainage of large districts this is the first matter to be attended to. A suitable water-course must be provided, into which all lesser drains may be discharged.

A ditch which is to be a water-course in ordinarily large sloughs, should be of greater dimensions than a cow-path, furrow or spade ditch. A narrow ditch, even if deep enough, will soon wash at the sides, causing sods and earth to fall in. These, with the growth of grass, will soon obstruct the ditch to such a degree that it will be worthless, unless the water flows rapidly enough to wash out all matter, or it is cleaned by hand-work. Besides this it will hold so little water that in every little freshet the land on each side will be flooded and injury done.

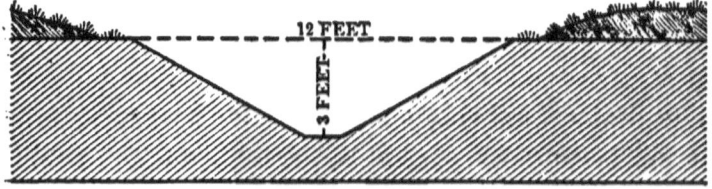

Fig. 5.—Proper Form for Open Ditch.

Figure 5 gives the cross-section of a ditch which will stand at the sides and can be easily kept clean. The ditch is twelve feet wide at the top and three feet deep. The sides slope 2 to 1, that is, one-half the width at the top is twice the depth. The earth should be taken 3 feet from the edge of the ditch, and should be smoothed and seeded to grass. A border of 10 feet should remain in grass. This will require two rods of land for the ditch, giving firm land for the banks. Such a ditch can be kept clear of weeds and long grass by mowing with a machine and burning the weeds in the bottom of the ditch. In making the ditch short turns should be avoided as much as possible, as these retard the flow, and occasion washing away on one side

of the ditch at the turn. The grade may be, in almost
all cases, uniform with the slope of the surface, as usually the inequalities are very slight in prairie sloughs.

The magnitude and expense of a ditch of this description at first often induces the farmer to substitute
some more easily constructed one, and thus cripple his
whole system of drainage. After a few years of trial
he will have reason to regret his half-way work, and
will take measures to correct it. When taken at a
dry season of the year, the ditch as above described
may be excavated quite rapidly with the help of a
road-plow and scraper.

TILE DRAINS.

The good effects of drainage previously mentioned
can not be brought about by a system of open drains,
only as such a system is constructed for the purpose of
affording sufficient outlets for under-drains. In observing the process of natural drainage as shown in
fig. 4, we see that such drainage is very slow, since it
depends upon the nature of the soil and the relation
of the contour of the sub-soil to the surface. Open
drains are simply an aid to natural drainage, acting
principally upon the upper six or eight inches of soil.
Deeper than this, the soil, during the spring-time, is
tough and compact, scarcely allowing the plowshare to
cut and turn it to the surface, because of its adhesive
nature. At the same time, a few inches of the surface
soil which has been surface-drained and acted upon by
the sun and air, will be friable. Later in the summer,
if the season is dry, the lower soil will be found partially dry, but generally it never becomes well drained
except at the surface. We must have ditches, but they
should be regarded only as necessary accessories to

under-drains, if we wish to realize their full benefit. A tile-drain, in order to accomplish its purpose perfectly, should possess the following requisites:

It should consist of pipes of sufficient size, laid at proper depths, to carry away all water which may come to them.

Each line should have a perfectly free outlet.

The pipes should have sufficient space between them at the ends to permit water to enter.

Each separate line should be laid on an incline, or series of inclines, of regular grade.

The tiles should be of good material and well-burned, in order to be a permanent improvement.

CHAPTER II.

ACTION OF DRAINS UPON THE SOIL.

How Tile Drains Affect the Soil—Temperature—Chemical Change
—Drought—Questions to be Considered Before Commencing to
Drain.

HOW WATER ENTERS A TILE DRAIN.

A correct understanding of this will help us to determine the best way to make the joints, and also to
locate the lines as regards their distance apart. The
tiles should have their ends joined as closely as the
inequalities arising from moulding and burning will
admit of. When this is done there will yet remain
sufficient space for the water to pass in or out, but
not enough to admit soil, except in the form of very
fine silt. At the bottom of the drain and nearly
on a level with either side of it, the earth is saturated
with water, that is, it can hold no more. The plane
forming the upper surface of this saturated earth is
called the water-table. Figure 6 shows a cross-section
of a drain, the curved line AB representing the water-
table, or line of saturation, the darker part of the figure
repesenting the saturated earth, and the lighter portion
above the water-table the drained soil. When rain
falls upon the surface it descends directly downward
by the force of gravity. When all the particles of the
drained soil contain all they will hold by absorption,
the water passes down until it reaches the saturated
soil, when, as it can go no further, it saturates the
lower portion of the drained soil, thus causing the
water-table to change its place and rise higher. As
the water-table rises, the water rises through the joints

of the tiles, and they being inclined, a flow begins and continues until the water-table recedes to the floor of the drain, when the flow ceases. It will be seen that the water-table will vary in height with the quantity of drainage water in the soil. When the water-table rises to the top of the drain, the tile will discharge a stream as large as its caliber. If the water-table rises higher than this, additional head is given and the ve-

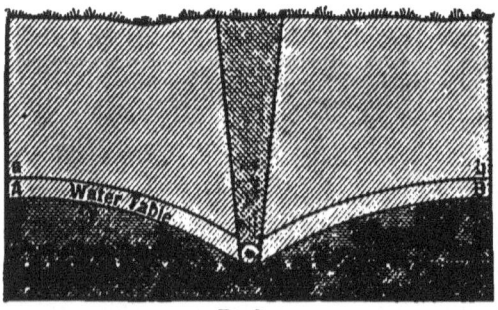

Fig. 6

locity of flow is increased, but the depth of drained soil is decreased. The fact that the tiles are porous does not increase the flow nor add to their draining properties. They would be as suitable for draining purposes if made of glass, or of glazed ware, as when made of porous clay, for they will be taxed to their full capacity by water flowing into the joints. The water-table does not extend on a level indefinitely either side of the drain, but rises as it recedes, the angle of rise varying with the nature of the soil. This fact will be alluded to again in the discussion of the distance apart of the drains.

HOW TILE DRAINS AFFECT THE SOIL.

Depth of Soil.—From what has been said before, it will be seen that the depth of the soil is increased by the action of tile-drains, since, were it not for the presence of the drain, when the water-table rises high, thus decreasing the depth of drained soil, it would remain so until the surplus water was carried off by slow natural drainage, in place of rapidly, as by the drain. All the soil acted upon by the drain is made similar to that at the surface. Air takes the place of the surplus water, so that a chemical action is begun. The inert soil matter is slowly changed into plant food, making the whole depth of drained soil the natural home for the roots of plants. It is often thought that the roots of farm plants penetrate but a few inches into the soil, and that if the surface is dry, rich and porous to a depth of ten inches, the plants have sufficient room for growth. Professor W. O. Atwater, in the *American Agriculturist*, gives the following on the "Extent and Depth of Roots":

"I have often been interested in noting the ideas most people have as to how far and how deep the roots of plants extend. The majority guess roots of grass and clover penetrate between five and ten inches, and are surprised to find that they reach several feet. I have some roots of timothy, clover, and other plants, dug from a very heavy clay soil, a good quality of brick clay, so compact and hard that a sharp knife, in cutting it, leaves a surface as smooth and shiny as it would cut on the end of a pine board. I have traced the roots of the timothy to a depth of two feet and four inches, and the clover three feet and two inches. A number of years ago a very intelligent German farmer named Schubart, made some very interesting observations upon the roots of plants as they grow in the field. An excavation five or six feet deep or more, was dug in the soil so as to leave a vertical wall. Against this wall a jet of water was played by means of a garden sprinkler; the earth was washed away, and the roots of the plants growing therein laid bare. The roots thus exposed in a field of rye, in one of beans, and in a bed of garden peas, presented the appearance of a mat or felt of white fibers, extending to a depth of about four feet.

"Roots of wheat sown September 26, and uncovered the 26th of April, had penetrated three and a half feet, and six weeks later about four feet, below the surface. In one case, in a light subsoil, wheat roots were found as deep as seven feet. The roots of the wheat in April constituted forty per cent. of the whole plant. Hon. John Stanton Gould, I believe it is, says that he 'has seen the roots of Indian corn extending seven feet downward,' and Prof. Johnson states that 'the roots of maize, which in a rich and tenacious earth extend but two or three feet, have been traced to a length of ten or even fifteen feet in a light, sandy soil.' Roots of clover, when growing in a rich, mellow soil, extend far, both laterally and vertically. Prof. Stockbridge 'washed out a root of common clover, one year old, growing in the alluvial soil near the Connecticut river, and found that it descended perpendicularly to the depth of eight feet.' Lucern roots are stated to reach a depth of twenty and even thirty feet. Alderman Mechi, in England, tells of a neighbor who 'dug a parsnip, which measured thirteen feet six inches in length, but was unfortunately broken at that depth.'"

It will be seen by this that maximum crops can not be expected until the soil is made light and porous to a sufficient depth to give the plant abundant nourishment. In prairie soil the depth which is so desirable can be obtained in no other way than by under-draining.

Temperature.—A warm soil is another effect of under-draining. When the soil becomes saturated and no means are provided for the removal of the water except by evaporation, no heat is absorbed by the soil until the water at the surface has been changed to vapor. In the summer the air is very much cooled by a shower of rain, because a certain amount of heat is required from the air and earth to convert a portion of the rainfall into vapor. The same change is necessary when the soil is saturated. If the rainfall is frequent but very little soil is warmed, all the heat of the sun being required to change the water at and near the surface into vapor. If this is true of the surface it is doubly true of soil several inches below the surface, for the water at the surface must be evaporated and the temperature of the soil raised before any warming process can go on in the lower portions of the soil. A drained soil has been found to be from six to ten degrees warmer at seven inches below the surface than an undrained soil at the same depth. This difference in the soil often gives the farmer a season which is from two to four weeks longer, besides giving quick and increased growth to plants.

Chemical Change.—Heat is an important chemical agent. When permitted to enter the soil with air, important changes are made. Vegetable matter, hitherto inert, becomes further decomposed and mingled with mineral matter, thus making the lower soil similar to that at the surface. Again, the ammonia furnished us by the rain is held in a drained soil and aids in this chemical work. In order to see that such changes are made, let a portion of clay or hard-pan be taken from a depth of three or four feet and exposed to the atmosphere. Instead of remaining compact and solid, it will gradually crumble and in time will become chemically changed. The same action goes on

2

when the earth is in place and the air is allowed to find its way to it, which it can not do until the surplus water is removed.

 Drought.—It is often asked "If draining makes a soil dry in a wet year will it not make it too dry in a dry time?" It has already been shown that a drained soil holds a large quantity of moisture by absorption. The soil being very much deepened, the roots of plants have access to the moisture contained in a much larger mass of soil than when undrained. Again, a soil is filled with capillary* tubes, which carry moisture to the surface, where it is quickly converted into vapor. If the surface is mellow and the whole depth of soil loose, the tubes are much larger, so that water is conveyed to the surface in much less quantities. Consequently, less moisture is lost by evaporation. Still further, in dry times the soil below the surface is much cooler than the air, hence, when air containing vapor is brought in contact with it, the vapor is condensed into water and absorbed by the particles of soil. In an undrained soil the surface is made compact by standing water, is baked by the sun when the water is evaporated, is compact below, giving little depth of soil for the plants. Moisture evaporates rapidly through the hard surface, and roots, having a comparatively

---

* The force called capillary attraction is of great importance to the cultivator of the soil. It is so called from the fact that it is most noticeable in very small tubes, called capillary tubes. It is the attraction which exists between a liquid and a solid when brought into contact. When small tubes are placed in water the liquid rises in them higher than the surface of the water, rising highest in the smallest tube and varying in height inversely as the diameters of the tubes. The soil contains an endless number of capillary tubes which communicate with each other. In hard and compact soils these tubes are much smaller than in loose and mellow ones, and according to the law, moisture is conveyed to the surface, where it is evaporated. If the surface be broken up by any means the capillary tubes are made larger to such an extent that but little moisture is conveyed by them to the surface to be evaporated.

small range, soon feel the ill effects of dry weather. Some soils are naturally very rich and porous, producing good crops when the spring rains are light enough to allow the soil to be worked, but it has been found that such soil produces much larger crops even in dry times when well drained. In short, thorough under-draining has been found to be a most efficient preventive of drought. It also makes better tillage possible, which in itself is a great advantage, and it makes all parts of the soil available for the use of useful crops.

QUESTIONS TO BE CONSIDERED BEFORE COMMENCING TO DRAIN.

There are many farmers who have poor crops and consequently find very little profit in farming, not because their land suffers from an excess of water in the soil, but by reason of poor management and poorer tillage of a soil which is naturally well drained. It is a weakness, particularly prevalent among western farmers, to put a little work on a great deal of land. The farmer should understand the value of labor upon the soil before he invests much capital in under-drainage. If the land which he has been cultivating has become filled with seeds of noxious plants until the crops consist of half grain and half weeds, let him turn his attention and capital to more thorough cultivation as the first thing to be done in order to make agriculture profitable. If his soil has become "furrow-trod" and produces a sickly crop, let him rotate his crops, plow in the fall, seed to grass and pasture until, by the best means known to agriculturists, the results of his labor show that he has succeeded. He may cheat the soil for a time by poor cultivation, but at last the farmer will suffer the full penalty for his treatment of his most profitable friend.

The first question that a farmer should consider before undertaking drainage, is this: Is my land, which is naturally drained, receiving such treatment at my hands as to give me maximum crops? Draining is expensive, and in order to obtain profitable returns from the outlay he must cultivate in such a way that the soil will bring him the largest possible crops. The more money there is invested in the land the greater necessity there is for making sure of a corresponding return. The gardener who is to pay two hundred or three hundred dollars an acre for land near some city which will afford him a good market for his products, should carefully take into account whether the crops which he intends to raise, the amount of which will depend upon his own skill, capital and adaptation of the soil, and the prices he can get for them, will justify the purchase. He very well knows that with poor crops as his only source of revenue he would soon be lost in a hopeless labyrinth of debt. It is much the same with the farmer in draining. A careful, industrious farmer can drain his land and realize a handsome return from the investment. A slovenly man, who will not cultivate well the land which nature has enriched and drained for him, need not expect to make draining pay.

Again, fashion rules the farmer to an extent which he is often unwilling to admit. A new implement has its run in a neighborhood as surely as a new pattern of goods among ladies of society. If one man builds a convenient and showy barn many of his neighbors will speedily follow suit, regardless of their means and needs. If neighbor A drains a field and thereby increases his crops, neighbor B, seeing that he has succeeded well, will forthwith begin draining one of his fields. This habit of imitation is not to be discouraged, for it is an important incentive to advancement in all things. A good example is worthy of imitation. But in the last case mentioned neighbor A may have given the matter careful thought, and have laid ·his plans with so much foresight, and carried them out with such thoroughness, that the results are highly satisfactory, while B, seeing only the results, hastens to obtain similar ones without the preparatory care. He, perhaps, does his work at random, and finds the profits much smaller than he was led to suppose. Each man should make the case his own. In connection with what his neighbor has done he should study his own soil and the natural facilities he has for thorough work.

Another matter to be seriously thought of is the expense. Draining, if done thoroughly, is expensive at the outset. Of course a small drain here and there on the farm does not involve much outlay. But prairie farms are usually wet to such an extent that outlets must be provided at considerable expense, which, in themselves, give no adequate return, and often an entire system of mains and sub-mains must be laid before much profit can be directly obtained. A survey should be made, giving the elevation and distances of different portions of the farm or field. From this an approximate estimate of the cost of a drain or system of drains can be made. However much or little is done the farmer should be prepared with the requisite knowledge and money to begin at the right place and do thorough work as far as he goes. It must be remembered that under-draining is not like building a fence, which may be moved if it is not in the right place, or which in a few years will rot down and require rebuilding, but it is a work which, if well done, will benefit the farm for future generations as well as the present.

CHAPTER III.

LEVELING AND LOCATING DRAINS.

The Outlet—Leveling—Level Notes—Leveling Instruments—Location of Drains—Staking and Leveling for Drains—Field Notes of Main Drain—Computing Grade and Depth—Determining and Adjusting Grades.

THE OUTLET.

The first and most important consideration in good drainage is the outlet. We may use tiles as large or as small as we please, and lay them as accurately as a railroad is graded, and even go to the expense of locating the lines ten feet apart, and yet if the outlet is not free, the drainage will not be successful in all respects. The lack of good natural outlets is perhaps the greatest difficulty that the western farmers have to surmount in draining flat prairie. To make underdraining successful, he must often deepen the watercourse by making open ditches, even larger and deeper than the one previously described. In order that a tile drain may discharge all the water that it is capable of carrying, the water must flow away with perfect freedom.

It is thought by many that some contrivance may be used by which the force of gravity may be circumvented, and the drain made to discharge its water, notwithstanding a faulty outlet. A box or barrel is sunk at the lower end of the drain, and the water made to rise twenty or thirty inches and then flow away through a shallow open ditch. The effect of this is shown in

fig. 7. We must remember that if water will enter at
the joints of the drain, it will flow out into the soil
just as readily if the stream in the tile is retarded or
entirely stopped. The water, as shown in the figure,
flows through the drain to its outlet A, where, being
held by the sides of the pit or box, the water must
rise to the level Bb before it can flow off. The soil
above the drain becomes saturated up to the line Bb,
thus leaving a portion of land near the outlet undrained, or only partially drained, according to the

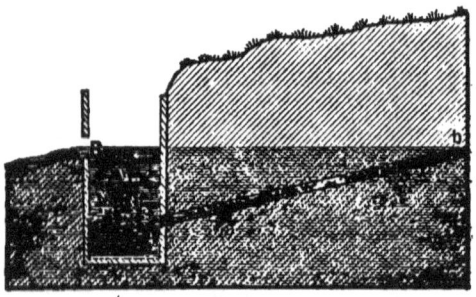

Fig. 7 - Faulty Outlet

relation between the height to which the water must
rise and the grade upon which the drain is laid. Nor
is this a small waste of land, especially where the
grade of the drain is slight. We will suppose that the
mouth of the drain is three feet below the surface, that
the water must rise twenty-four inches before it can
flow off, and that the grade of the drain is four inches
in one hundred feet. The drain will now back water
for six hundred feet, and will greatly injure the land
for a distance of three hundred feet above the outlet.
When the grade of the drain is great near the outlet,
the damage is not so serious. When the soil above
the drain becomes saturated from any cause, as in the
case of an obstructed outlet, the combined weight of

soil and water above the drain causes the tile to act like a continuous iron pipe, the head of water above forcing the water through the pipe to a proportionate height at the outlet. This, however, is not drainage. The drain must have a free outlet, and the tiles should never be quite full in order to give perfect drainage to all of the soil through which the drain runs. The proper construction of the outlet will be taken up under the subject "Location and Construction of Drains."

LEVELING.

Unless inaccurate and in the end costly guess-work is depended upon in draining, we must use some accurate method of obtaining the difference of elevation of various points on the farm or field upon which we wish to operate. Leveling can be done in various ways, but in whatever way it is done the principle remains the same. We first secure a horizontal line of sight, and, having a rod graduated to feet and fractions of a foot, we hold the rod upon the point whose elevation we wish to determine, and find at what mark the horizontal line intersects the rod. This is called a rod-reading. The rod is then moved to another point and a reading is again taken. The difference of the readings is the difference of elevation of the two points upon which the rod was placed. If we wish to take a system of levels, that is, find how much higher or lower different points are than the starting point, the process becomes more complicated, for the reason that the instrument must be moved and the levels all referred to the same datum. The following method of operating and keeping the notes is general, and will apply to whatever kind of instrument is used to obtain the horizontal line.

For convenience in leveling for drainage purposes, we begin at the place which we consider the lowest point upon the farm or field, and make a preliminary level survey in order to find the elevation of the lowest

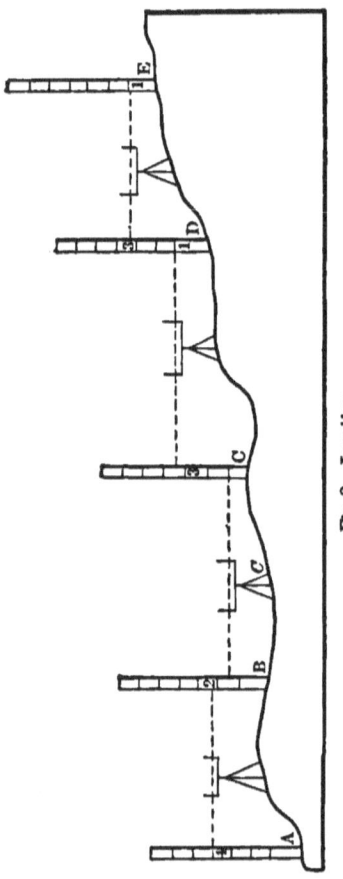

Fig. 8 - Leveling

portions of land requiring drainage, and the distance of such places from the common outlet. We assume the starting or outlet point at the surface of the ground to be 100 feet above an imaginary plane below called the *datum plane* or *datum*. Place the instrument at some convenient distance from this point (the distance will depend upon the power and accuracy of the instrument), take a reading at the point A, Fig. 8, which we will assume for illustration to be four feet; add this to the assumed elevation of A, and we have 104 feet, which is the height of the line of sight or of the instrument above *da-tum*. Now take the rod to B and take a reading which we will assume to be two feet. Subtract this reading from the height of the instrument and we have 102 as the elevation of the point B. Change the instrument to some place beyond B, as at C. Take another reading at B, called a backsight, or commonly a plus sight, which we will suppose is

1.5 feet. Add this to the elevation of B for the
height of the instrument in its new position, which is
103.5 feet. Take a reading at C, which is one foot.
Let these operations be repeated until the elevation of
all points desired is found. Observe that at every
change of the instrument a back-sight must always be
taken upon the last point at which a reading was taken,
and its reading added to 'the elevation of that point
for a new height of the instrument. Also subtract
every fore-sight reading from the height of the instrument, to obtain the elevation of that point. The notes
should be kept as indicated below:

LEVEL NOTES.

Stations.	Distance.	Back-sight or sight.	Height of Instrument.	Fore-sight or sight.	Eleva- tion.
A................................		4.00	104.00	......	100.00
B ....400 ft.		1.50	103.50	2.00	102.00
C..........................500 ft.		3.00	105.50	1.00	102.50
D ....525 ft.		3.00	107.50	1.00	104.50
E ....375 ft.		......	....	0.90	106.60

NOTE.—The *datum* height 100 is used to avoid minus quantities.
If the elevation of some point should be 98, it would indicate that
such a point is two feet lower than the starting point.

The distance between the points should be measured
and recorded in the notes. A rude sketch of the lines
and stations may be made. From this preliminary
survey the farmer can ascertain what fall he has in
given distances, and how he had best lay out his drains.
In short, he can tell what it is possible for him to do.

LEVELING INSTRUMENTS.

There are many kinds of leveling instruments within
reach of the farmer. Perhaps the ordinary water level
is as easily made and as efficient as any. This may be
made in the following manner: Get the tinner to make

a tin tube one inch in diameter and four feet long, turned up at the ends two or three inches, as shown in fig. 9. Insert in either end phials one-half inch in diameter, and long enough to allow about two inches above the tin, having previously knocked out the bottoms, that there may be free communication between them. The phials may be fastened in place by plaster of paris. Clamp the tube securely to a block of wood and place it upon a tripod, so that it may turn readily upon a pivot. The legs of the tripod may be hinged upon the head. This will add to its convenience. To use it, pour water in the tube until it is nearly filled, and adjust the tube so that the water may be easily seen in the glasses. The water having been previously colored by carmine, the height of water in the two phials will form a level or horizontal line. The eye should be placed at a distance of three feet from the tube, and raised or lowered until the line of sight prolonged will coincide with the height of the water in the tubes. An assistant should move the target upon the rod until a signal from the one at the level indicates that the target marks the intersection of the level line with the rod. A good eye-sighting along this line can, with considerable care, obtain very good results. This is a slow method for long distances, for the eye can not prolong the line indicated by the water in the phials to any great distance. When the level is to be moved, the phials may be corked until it is nearly adjusted again, when the corks should be removed.

A carpenter's level having sights upon it, and made to turn upon a pivot on a tripod, is another kind of level with which fair work may be done. Whenever a spirit level is used, it should be examined to see if it is in adjustment.   If not, correct work can not be expected to be done.   The advantage which the water level has, over the spirit level is that the water always indicates a horizontal line, while the spirit level may be out of adjustment to such a degree as to be unreliable.   The sights put upon the spirit level require accurate construction, or they will not give correct results.   Either level accurately constructed is reliable in itself, but it will be seen that the former is more easily made so than the latter.   Whichever instrument

THE COMSTOCK LEVEL.

is used, no little care is required on the part of the beginner if he wishes correct work.   He should first see how correctly his instrument will work, by finding the elevation of the same point from different positions of his instrument.   If the results disagree to any extent, he should try to find whether the error is in his work or in his instrument.

It is often the case that the water level is not expeditious enough in its work, and the engineer's level is too complicated and high-priced for the farmer.   To meet this want the Comstock Level, shown in the above figure, seems well adapted.   It is manufactured by

William T. Comstock, 6 Astor Place, New York.
The instrument is made of brass, lacquered so that it
will not tarnish.   It consists of a sighting tube A A,
fourteen inches long, having a pin-hole through one
end, through which to sight, and adjustable cross-wires
at the other.   The tube is mounted upon Ys Y Y', and
can be taken out and reversed upon them for adjustment.   The Ys are mounted upon a circle, which
moves within another circle, as shown at C C'.   The
inner circle contains the level bulb, which can also be
adjusted by a small screw beneath the plate.   One
quadrant of the outer circle is graduated to single degrees, and the inner one marked at intervals of 45°, so
that it may be used in laying off angles for buildings
and other similar work.   The instrument is leveled up
by means of the thumb-screws S S, which rest upon
the tripod-head B.   A plumb-bob with its line passing through the center of the instrument, permits its
center to be set over any desired point.   The instrument mounted and in adjustment is thus made very
convenient and well adapted to the purpose for which
it is made.   The level can be turned to any point of
the compass, and its parts are all arranged for adjustment.

The engineer's spirit-level is the most accurate instrument, and the one with which the most rapid work
can be done.   The farmer, by giving considerable study
to the subject of leveling, and using a great deal of
care, with common instruments can do work that will
be sufficiently accurate for ordinary farm drainage, but
he will find it, as a general thing, to be a matter of
economy to employ a surveyor with accurate instruments, whose experience and reputation in their use
will be a guarantee for correct results.

Mains.—Having found the difference of elevation of various portions of the land to be drained, we are prepared to fix upon the lines for the main drains. This is a work which, in many places, gives opportunity for the exercise of much skill in the use of knowledge pertaining to drainage. It will be assumed that sufficient level-notes have been taken, and distances measured, to determine the fall per 100 feet between the particular spot to be drained and its nearest outlet, or, if the land is nearly flat, the amount of slope it has in any direction.

The first knowledge that the farmer should avail himself of is, that which he can obtain by observation in the spring of the year, when the soil is saturated with water. At such times water will be found standing above the surface in hollows or basins in the land, and also on flats which seem as high as the surrounding surface. Mark these places and determine, if possible, whether the water is held by a clay sub-soil, as shown in fig. 4, or by the quantity of water retained in the soil at lower portions of the field. In the first case the natural drainage will be very slow, even though the elevation be sufficient; while in the latter, the natural drainage will go on rapidly if the surplus water is removed from the lower portions of the field, thereby giving the water an outlet through the soil. If the whole field seems nearly flat, see if there are not some spots which are wetter than others, though the contour of the surface does not indicate it. Upon examination it may be found that the cause of this is with the sub-soil, as before noticed, or with the soil itself, it being made up largely of clay and more retentive of water.

The bearing of these observations on the location of mains is this: There are places, such as have been mentioned, which must be drained by a system of branch drains. The nearer the mains can be brought to these the less will be the expense of the branches and the more effectual will be their action. By these observations the farmer has an accurate method of finding the lowest places through which all main lines of drains should pass. The variation in the course of the main to suit particular

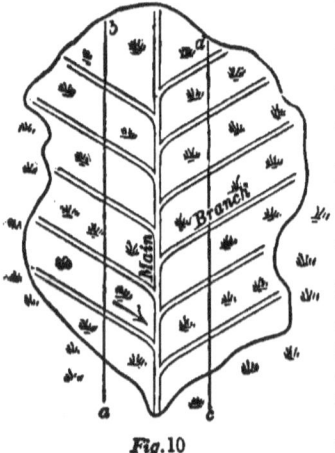

Fig. 10

cases is often precluded by the slope of the surface, and also by the extra expense a longer main would incur.

The general rule for the location of mains is to let them follow the lowest land, or course of natural drainage. The surface then slopes towards the drain, making both natural and artificial drainage easy. We might say here that in all cases we should try and take every advantage that nature has given us in this work, for artificial drainage is only completing the work which nature has begun.

There are cases which require us to make exceptions to the general rule just given. First, the drain should be as free from angles and short turns as possible. In other words, it should be laid on a straight line, or a series of straight lines, connected by long curves. A few words in explanation of the advantage of straight lines and easy curves will convince the reader of their importance. There is a certain number of feet or inches of fall that can be used in a given distance. The shorter we can make that distance by cutting across angles the greater the fall per 100 feet we can get, and consequent greater velocity of flow and discharge of the drain. When the total fall is slight, as is very often the case on land which suffers most severely for want of drainage, every thing that can be done to increase the velocity of flow is of prime importance.

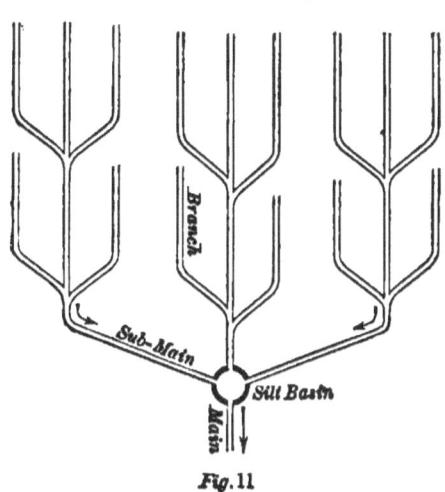

Fig. 11

Again, mains must be made of larger tiles than any other of the lines, and so cost much more per foot. Crooks in the line increase its length and consequent expense.

Short curves decrease the velocity of flow, so that, if we wish to have a uniform velocity in all parts of the drain, the grade must be increased at the bend, and as a consequence, the grade of the entire line must be lessened.

It will be seen that there are many things peculiar to each location to be considered in determining the

3

proper course for mains. In making the line shorter in order to lessen the expense and increase the fall, we may by a deep cut, made to avoid some turn, increase the cost more than all we save, or by so doing we may fail to drain some land through which the drain should pass.

It is only by carefully weighing all those things which enter into the expense and efficiency of the work that the farmer or drainage engineer can arrive at the most desirable plan.

It must always be borne in mind that in small ponds, drained by a single line of tile, the drain should pass entirely through the pond, and thence to the outlet, instead of beginning at the edge of the pond as in the case of an open ditch. The reason for this is evident when we remember that water from the land on either side of the drain enters it through the joints of the tile, while the land at the end of the drain is drained but very little.

Sub-mains and Branches.—It is often the case that a single line of tiles laid through a flat, basin or hollow will afford a sufficient drainage for the purpose of the farmer. When we wish for the thorough drainage of flats, ponds or swamps, we must have mains to give an outlet for the water when collected, and a system of sub-mains and branches to collect the water from the soil and discharge it into the mains. There are different systems of laying out branches, the value of each depending upon the area to be drained.

Figure 10 shows a system commonly used, but as a general thing it is not to be commended. It consists of parallel branch drains discharging into the main, often at right angles to it. If the main is of proper size it will of itself drain the soil for a distance of from

forty to fifty feet on either side.   It will be seen that
the portion of the branches between the lines a b and
c d are then superfluous, the main alone being sufficient to drain that amount of land.

A better system for large areas is shown in fig. 11.
Parallel sub-mains discharge into a silt basin, and
branches parallel to these discharged into the sub-mains
with but little waste of drains, each drain acting upon
its own area of soil.

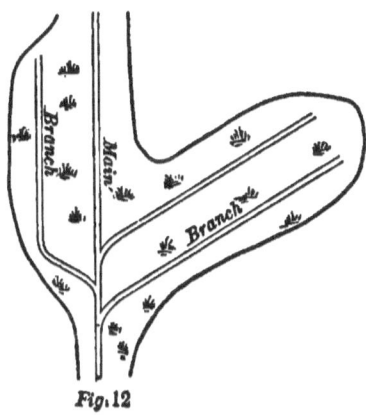

Fig. 12

In fig. 12 is shown an adaptation of both systems to
suit the land.   Upon the right is a narrow piece of flat
land, extending some distance from the main, which
can be drained most economically by two parallel
branches running at a proper angle to the main.   At
the left of the main is a flat extending lengthways of
the main, but wider than can be drained by it, hence a
small branch is laid out parallel to the main.

These examples are, perhaps, sufficient to show the
way in which branch drains are to be adapted to accomplish the work desired.   Their distance apart and
depth will be treated of under their appropriate heads.

The junction of all branch drains with mains and

sub-mains should be at such an angle that the pipe wil
discharge as nearly as possible in the direction of th
current of the main and larger stream. Where it i
necessary to have the drains connect at right or obtus
angles, the junction tiles should be curved in the di
rection of the current into which it discharges. It wil
be observed that in the figures given all angles ar
avoided. When a change of direction is desired curve
are used. The reasons for this will be discussed mor
fully further on.

We wish to urge upon all who are about to under
take drainage that the application of correct principle
to practice is what is most needed. We can no
always fully carry out a correct theory in practice, bu
the nearer we come to it the better will be our work

STAKING AND LEVELING FOR DRAINS.

The next step is to lay out the drains and properl
prepare them for their actual construction. Thi
should not be done by guess work, either by the farme
or professional ditcher. Draining on prairie land is
very different thing from draining on hillsides, or nea
the bank of some lake or creek. We can not afford t
waste time, money and strength trying experiments i
draining, when the success of the work may be know
before the ditch is begun or a single tile laid. Ther
are several methods of preparing the drains for th
ditcher, and it is probable that the one that is bes
understood by the operator will seem to him the mos
desirable. The method we shall describe we think ha
the advantage of being applicable to every case, an
easy to work from in constructing the drain.

We shall attempt such a minute description of i
that any one, whether he possesses an engineer's leve

water level, or any other means of obtaining a horizontal line, can stake out and level the drains.

In all the operations of staking out, leveling, grading and laying drains, we begin at the outlet. This is the base upon which to found our calculations. Having previously determined the position of the outlet, and the course of the drains, from observations made in the early spring, and the preliminary levels before mentioned, we prepare stakes of two kinds with which to stake out the lines. One of these is called a *grade peg*, and should be about one inch square in section, and eight inches long. The other is called a *guide stake*, and is best when made of boards one inch thick, two inches wide and two feet long. The upper four inches of this stake should be planed on one side sufficiently smooth to receive and hold a pencil mark. An idea of these stakes and their use may be obtained from fig. 13. Knowing about the

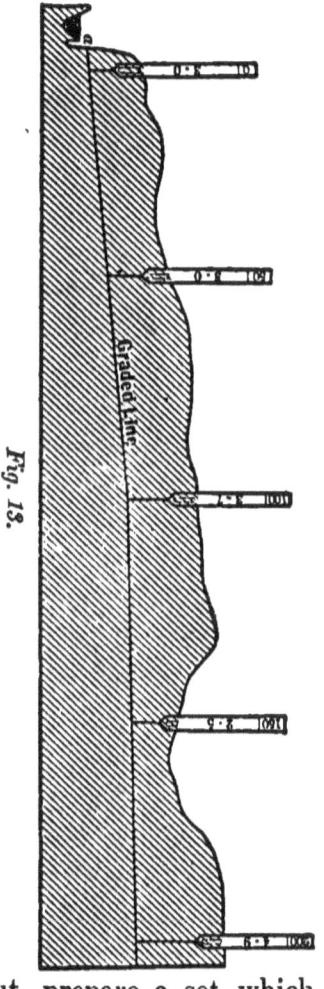

length of drain to be laid out, prepare a set, which consists of a grade peg and guide stake for each fifty feet of length. We also need a measuring chain or a tape line graduated to feet, and a hatchet with which to drive the stakes.

Begin at the outlet of the main and drive a grade peg about one foot to the right of the center line of the drain, and as near the end as is desirable. At the right of the peg and about five inches from it, set a guide stake, near the top of which mark O with a heavy lead pencil (see figure). Measure a length of fifty feet from the O stake and set a grade peg and accompanying guide stake, as in the first case, and mark the figure 50 upon the guide stake. The upper marks upon the guides indicate the number of feet of length from the outlet. The grade pegs should be placed as near to the edge of the proposed ditch as they can and remain firm after the ditch is dug. There is no reason why the stakes should be set fifty feet apart rather than any other distance, except that in practice it seems to be the most convenient to work from. Let the whole main line be staked out in this manner, numbering the stakes O 50, 100, 150, etc. Where there is a curve in the ditch let the chain or tape follow the line, that we may know the exact number of tiles required.

Another thing to be noticed at the time the main is staked out is, where the sub-mains and branches enter. Wherever a junction is made a grade peg and guide should be placed. This peg marks the outlet of another drain, and should be marked O, and its name in addition to its number, on the main line. An example of this may be seen at stake 150 in fig. 13.

Some system of designating drains is needed where there are many, in order that the notes may be kept without confusion, and correspond with the plat which is made after the drains are staked out. The main is known simply as the main. A sub-main is a prominent branch of the main, which has branches of its own. We call the sub-mains branch A, branch B, etc., in order as they are laid out, counting from the outlet of the main. Branches of the sub-mains are designated as No. 1, 2, etc., of A; No. 1, 2, etc., of B; and so on, numbering in order from the junction of the sub-main with the main. The guide stakes on each branch are numbered from its junction towards its source.

Having staked out our lines, we next take our leveling instrument, however simple it may be, and find the elevation of each peg by setting the level rod on the peg, which should be driven down even with the surface, and taking a reading in the manner described under the head of leveling. The pegs being driven to the surface of the ground, their elevation above the *datum* plane will truly represent the surface of the ground along the line of the drain. Call the elevation of the peg at the outlet 100 feet. Take another level where the outlet tile will be placed (see *a*, fig. 13), and be sure that there is sufficient fall below it to carry away the water as fast as it is discharged. This last level will give the elevation of the bottom of the tile at the outlet, or what we call the grade line at O stake.

The method of leveling and keeping the notes is the same as previously described, except that it is more complete. There are added to the notes before described, three columns for computing the grades and depth. The following is given merely as an example to show the method of keeping the notes. Figure 13 gives a section for the purpose of showing how the stakes are set and marked, both for depth of ditch and length of line, but it is not a true profile from the notes here given. This will be referred to again.

FIELD NOTES OF MAIN DRAIN.

Sta.	+S.	H. of I.	—S.	Eleva-tion.	Grade Line.	D'pth	Dep. in ft. & in.	Remarks.
O	645	106.45	......	100.00	97.25	2.75	2—9	Grade 50 to 100, or 6 inches to 100 feet.
A	......	....	9.20	97.25	....	......	....	A as low as outlet tile can be laid.
50	......	....	6.32	100.13	97.50	2.63	2—7⅝	
100	......	....	5.92	100.53	97.75	2.78	2—9⅜	
150	......	....	5.61	100.84	98.00	2.84	2—10⅛	
200	......	....	5.12	101.33	98.25	3.08	3—1	
250	7.10	108.94	4.61	101.84	98.50	3.34	3—4⅛	
Br. A	......	....	......	....	......	......	....	
300	......	....	6.72	102.22	98.75	3.47	3—5¾	Branch A enters at 300.
*	......	....	......	....	*	......	....	*Change grade to 32—100, or 7⅞ inches to 100 feet.
350	......	....	6.11	102.83	98.91	3.92	3—11⅛	
400	......	....	6.72	102.22	99.07	3.15	3—1⅞	
450	......	....	7.21	101.73	99.23	2.50	2—6	450 middle of sag or pond.
500	......	....	7.11	101.83	99.39	2.44	2—5⅜	

Each drain has a set of notes similar to the above. The notes for branch A would be headed branch A from stake 300 of main. A branch of A would be headed No. 1 from 225 of A. By this system of notes everything pertaining to the survey is kept distinct, and may be referred to at any time. We will next take up the method of computing grade and finding the depth.

COMPUTING GRADE AND DEPTH.

It will be noticed that in the "Field Notes" given, there are three columns of figures more than were given under the head of "Leveling." The first of these is headed "Grade Line," so called because it contains the elevation above *datum* of points on the line upon which it is proposed to lay the drain. These points are below and opposite every "grade peg" whose elevation has been found and recorded in the column of "Elevations." When connected by a line they form the "Grade Line," or line which determines the position of the drain with respect to its depth.

The second additional column, headed "Depth of Cut," or simply "Depth," gives the distances from the grade pegs to the grade line, or the depth of cut to be made, measured from the pegs. The numbers in this column are feet and decimals of a foot, as these are much simpler for computation. The last column of the three is the same as the preceding one, except that the numbers are feet and inches in place of feet and decimals of a foot. This change from decimals of a foot to inches is for convenience in digging the ditch, as the common measuring square is the most convenient and most commonly used in laying off a measuring gauge. If the leveling rod used in the work is graduated to feet and inches, only one column will be necessary, as the depth will be given in the same units. Using the scale of feet and inches, however, makes the computations much more difficult than they are when the decimal scale is used. The column headed "Remarks" contains comments upon the grade, location, etc., which are often a very valuable part of the notes.

We have briefly explained what these columns contain. We will now notice the method by which they are obtained. The starting point of the grade line was ascertained when we found the place for the outlet of the drain, shown at a fig. 13, and recorded in the notes as 97.25. We have found by previous leveling that there is sufficient fall below to allow the water to flow away freely, if we locate the outlet of the line at this point. Now, ascertain by the notes the fall we have in the given distance (the principles involved in this will be discussed hereafter), and divide this by the number of hundred feet, and we have the grade, or fall per 100 feet. Suppose that this, as given in the first of the notes before referred to, is .50 feet, or .25 for each 50 feet of length. Add .25 to 97.25, and this will give us the grade line elevation at stake 50, or 97.50; then to this result again add .25, and we have the grade line elevation at stake 100, or 97.75, and so on—continuing to add .25 to each grade line elevation to obtain the succeeding one, until, perhaps, it becomes necessary to change the grade, when the new grade, or fall per 50 feet (which in the notes given is .16), is substituted for the .25 in the successive additions, and the process continues as before.

. To obtain the next column, subtract each grade line elevation from the elevation of the peg at the same station, or subtract each number in the grade line column from the corresponding number on the same horizontal line in the elevation column. The result will be the depth of ditch to be dug at each peg. Thus, by subtracting 97.50 the grade line elevation at station 50 from 100.13, the elevation of the grade peg at the same station, we have 2.63 as the depth of the cut. The decimals in this column are then changed to inches, and the depth in feet and inches written, for convenience, in a separate column, thus 2.63 feet becomes in the next column 2 feet 7⅝ inches. The numbers in this column headed "Depth in Feet and Inches" are now marked upon the stakes at the respective stations, as shown in fig. 13. The stakes now show two sets of numbers. The upper one indicates the distance from the outlet; the lower one shows the depth at which the tiles should be laid, measuring from the top of the grade peg. This is the mechanical opertion of computing grades, and we trust that sufficient explanation has been given to make the methods plain. There are, however, important principles which should govern this work in order to insure economical results.

DETERMINING AND ADJUSTING GRADES.

We will suppose that the desired drains have been staked out and leveled, and we have all of the necessary leveling notes in our book. Determining grades requires much skill if it is done rapidly and in the best manner. It at least requires careful thought, and the inexperienced will find that a profile made from the notes will greatly aid in the work.

The question of how little grade a drain may have and work successfully may be answered by asking, How accurately will the drain be laid? We know of drains laid on a grade of one inch to 100 feet. Such drains must be laid with the greatest possible precision, or they will fail. Drains laid on a grade of two and three inches have proved eminently satisfactory when the work has been well done. Were it possible to avoid it, we should never lay a drain on a less grade than two inches in 100 feet. But there are many places, where, if we drain at all, under the present system of things, we must lay drains upon a less grade, and when it is necessary to do so, too much pains can not be taken in finding a correct grade, and laying the tile with precision.

Entering more into the details of deciding upon grades, for the sake of giving a clear idea of every step, we will suppose that the outlet of the drain is to be 3 feet deep, and the length of the drain 700 feet. By subtracting the elevation at the outlet from the elevation at the source, as found in the notes, we find that there is a fall of 28 inches in that distance. Dividing the number of inches of fall by the number of hundred feet, we have 4 inches as the grade per 100 feet. If the drain is run on this grade it will then be as deep at the upper end as at the outlet. If we wish to have the drain 3 feet deep at the outlet, and only 2 feet 5 inches deep at the upper end, we shall have 7 inches more fall, which will be available to use in the grade of the drain, making a fall of 35 inches, or 5 inches per 100 feet. Again, supposing we wish to have the outlet only 2 feet deep, and the upper end 3 feet deep, the fall which can be used in the grade of the drain will then be decreased 12 inches, which leaves us only 16 inches total fall, or 2⅔ inches per 100 feet. Again, taking the same example, we have the same total fall in the whole distance, but there is a rise of ground somewhere between the outlet and the upper end, as may be seen in fig. 13, through which, if the drain were laid on a uniform grade, as noticed in the first examples, much deep digging would be required, which may be avoided by using two or more grades. For example, let us lay out the first 300 feet from the outlet on a grade of 6 inches per 100 feet. We have now used 18 inches of the fall, and if we wish the drain to be the same depth at the source as at the outlet, we have only 10 inches fall left for the remaining 400 feet, or 2½ inches per 100 feet. This gives a grade that may be relied upon to do good work, and will not be as expensive to construct. It is often the case that there is not sufficient fall to permit the use of two or more grades on the same line; in such cases the deep cut can not be avoided.

We might go on multiplying examples, but those already explained may be sufficient to show the way any practicable system of grades may be made out. Each new field upon which drainage is undertaken will present examples peculiar to itself, which will afford abundant opportunity for the exercise of skill on the part of the farmer or engineer.

It is desirable that the grade of the drain increase toward the outlet, rather than the opposite, though this can not always be done. When a branch enters another drain, it is best to have an outfall of an inch or two, in addition to the grade of the drain. This overcomes the resistance offered by the bend at the junction. If the grade is very slight, it is not best to sacrifice any of it in this way, but let the branch enter on the regular grade. At every bend in the line the additional friction causes a decrease in the velocity of the flow, and it is best to increase the grade at that point, especially if the bend is somewhat short.

These few suggestions may be enough to enable any one to adapt his work to the case in hand. The depth of drains, which will be considered soon, will often have much to do with determining grades. No absolute rule can be given; but uniformity should, as much as possible, be aimed at. Carelessly laid out grades are a fruitful source of trouble with drains. If the water alternately flows rapidly, and then almost stands still, the tile is at one place full, and at another only part full. It will be seen that the tile will discharge only part of what it is capable of doing and would do if it were laid on a carefully arranged grade. A careful survey and corresponding grade will often add one-half to the efficiency of the drain.

CHAPTER IV.

DEPTH AND SIZE OF DRAINS.

Silt Basins—Depth and Distance Apart of Drains—Sizes of Tile—
Kind of Tile—Concrete Tile.

SILT BASINS.

The silt basin is often a valuable auxiliary to a system of drains, but it is not used as much as it would

Fig. 14 - *Silt Basin*

be if its advantages were better understood. It may be described as a small well, placed either in the line of a single drain, or at the junction of several drains, and serves several different purposes. Figure 14 gives an idea of the construction and use of the silt basin. It may be built of brick, stone, or plank, and may vary in diameter from twelve to twenty-four inches, according to the use for which it is intended. There should be a depth of twelve inches below the tile for the deposit of muddy matter. In the figure given it is shown two feet in diameter, constructed of brick, with a stone foundation. In draining it is often desirable that several sub-mains or branches should join at one place, and there unite in one line as an outlet to the whole system. This is the use of the basin, as shown in the figure here given. It permits us to unite several drains entering at different angles, without the objectionable feature of short turns, which we have before noticed. To facilitate the action of the drains, the outlet of the basin should be a few inches lower than the outlets of the lines of tile entering it.

Another advantage is, that the fine earth, or "silt," as it is called, which finds its way into the tile and is carried along with the drainage water, is permitted to settle in the basin, instead of being carried on by the current, to lodge in some portion of the drain where a turn is made, or where the velocity is decreased by a less grade. The basin should have a cover, which may be removed and the silt taken out before it impedes the flow of water through the tiles.

Another use of the silt basin is to prevent the silt from obstructing the drain in cases where the grade suddenly changes from a steep grade to one considerably less. This retards the flow, which causes the silt coming from the upper part of the drain to be deposited at the point where the change to a less grade is made. Here is where the basin should be placed, in order that the silt may be intercepted and removed when the lower portion of the basin becomes full. For this purpose the diameter of the basin may be much less than for the purpose of collecting the water of several drains.

In the ordinary drainage upon western farms, there

is but little necessity for the construction of basins for
the purpose of simply collecting silt, for there is usually not enough difference in the grade to cause any
alarm on that account. Yet near streams which break
the land up into alternate steep slopes and flat bottoms,
they are sometimes a necessity. In long mains, however, it is best to locate silt basins at various places
along the line for the purpose of watching the action
of the drain, and to see that its several sections are in
perfect condition.

Where the soil consists of loam on a firm clay sub-
soil, there is very little and sometimes no deposit of
silt after the drain has been in operation a few weeks.
There are many sub-soils, even in prairie lands, which
contain streaks of sandy material, which, for some
time after the construction of the drain, will find its
way into the tiles. It will be seen that the provisions
made for the interception of silt must be regulated by
the kind of material in the soil through which the
drain runs.

We can not urge too strongly the use of the silt basin for the purpose of collecting the water of several
drains into one, and thence conveying it to the ulti-
mate outlet. In the system of laying out drains, described in a former paper, and shown in fig. 11, the
use and importance of the silt basin is shown. A ju-
dicious use of the silt basin for the several purposes
for which it is intended, will greatly increase the efficiency of any system of drains.

DEPTH AND DISTANCE APART OF DRAINS.

Depth.—So intimately are the subjects of depth of
drains and their distance apart connected, that we can
not fix upon one without taking into account the other.

The first question which should be answered is in reference to the depth which we wish to drain the soil. What is the most suitable depth for the soil we have, and the purpose for which it is to be drained, taking into account the cost of drains at different depths, and the comparative advantages to be derived from them? The drains must at least be placed deep enough to receive no injury from frost during the winter. This is about two feet, though drains much nearer the surface than this have done good work for some time, but can not be regarded as safe. How much deeper than this we had better go depends upon several facts and principles, to which we hope the reader will give attention, for in this, as in many other matters of drainage, no absolute rule can be laid down and mechanically obeyed.

Many farmers have a mistaken idea in thinking that the removal of surface water sufficiently to fit the soil for plowing in the spring, and comfortable tillage during the summer, is the sole object of drainage. The advantages of a deep soil, and the use made of it by growing crops, have been explained in previous chapters. If we wish a deep soil, it is evident we must remove the surplus water and admit the air to the depth to which we desire the roots of the plants to penetrate and receive nourishment. We hear many arguments in favor of very shallow cultivation of growing crops, on the ground that the roots at the surface will be cut in pieces and so deprive the plant of nourishment. This argument will apply only to undrained or shallow drained land, where it will be found that the great bulk of the roots lie near the surface, only a few penetrating the undrained soil.

4

In order to obtain a given depth of drained soil, let us notice a few things connected with the action of drains which were merely hinted at in the explanation of fig. 6. In fig. 15, we have a section of three drains T T T. It was stated under this heading, "How Water Enters a Tile-Drain," that the line of saturation on either side of the drain is a curved line, and that this line, or, more properly speaking, the water-table, varies in height according as the soil is completely drained or only partially so. The line of saturation may be as represented by the line *bb*, or it may descend much more sharply towards the drain, as is represented by *cc*. The difference in these curves is caused by the nature of the soil. In the first case the soil is light and easily penetrated by water, and but little resistance to the flow toward the drain is offered by the soil. In the second case the soil may be clay, very retentive of water, and in its nature will

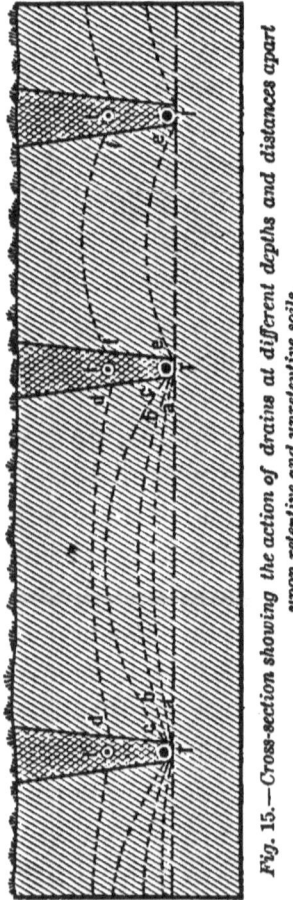

Fig. 15.—Cross-section showing the action of drains at different depths and distances apart upon retentive and unretentive soils.

not allow water to flow towards the drains except at a steeper incline. In this case a portion of soil between the lines of tile is left undrained, which, in the case of a more permeable soil, would be well drained. Some writers assert that there is no lateral flow of water through the soil towards the drains, but a few experiments and careful observation will prove the contrary, and the truth of the reasoning herein given.

It will be seen that the kind of soil to be drained governs the depth at which the tile should be laid, and also the proximity of the lines to each other, when we consider the thoroughness of the drainage.

The advantages of deep drainage are, first, a greater amount of soil is made available to crops, and fewer ill effects are felt from drought; second, there is room for more water in the soil in times of heavy rainfall, so that water may rise considerably above the drains for a short time without seriously affecting the crops. Suppose the drains are located at *ttt* (see figure). The soil becomes drained no deeper than the floor of the tile. In time of heavy rain the water can not pass off as fast as it falls, and of course saturates the earth much above the drains, and often to the surface. In this case the tiles must be much larger, so as to carry off the water nearly as fast as it falls, if we wish to keep depth enough of drained soil so that no injury for the time being may be done to the crops. This will many times account for the cry often made, "My tiles are too small." The same tile placed deeper, thereby giving a larger reservoir in which to collect drainage water in times of heavy rains, would often remove the difficulty.

We have thus far been general in the discussion of depth. The question will be asked, "What particular depth is most preferable, all things considered?" If we are careful to lay out grades to the best advantage, our depth will vary much with the inequalities of the surface. From the experience of many, it has been found that a depth of from 3 to 3½ feet in prairie soil is most desirable. It will be found that some portions will be laid four feet deep and others only three feet, or even less, if we aim at a general depth of 3½ feet. The expense of digging the ditches for four-feet drains is much greater than for three-feet drains, so that for general purposes of farm drainage the above instructions may be regarded as the best that can be given. It is not always possible, however, to obtain the desired depth, because of the shallow outlets which farmers are sometimes obliged to use.

Distances Apart.—According to the principles already noticed, drains in a retentive clay soil must be placed nearer together than in ordinary vegetable loam, if we wish to drain all the land between them. Even then the water-table will not recede so near to the floor of the drain as when the water percolates more freely and rapidly through the soil. In our experience, drains placed 100 feet apart in our loamy soil, and 3½ feet deep, will thoroughly drain the land where the surface is ordinarily flat. It has been found that so easily and rapidly does our soil drain, there is no necessity for such close proximity of drains as is used in the East.

If, however, the soil is very retentive, especially near the surface, a distance of from 50 to 75 feet may be required to give thorough drainage.

Let us now sum up the subject of Depth and Distance Apart of Drains. The line *aa*, fig. 15, may represent the water-table when the soil is not retentive to any great extent, and is completely drained. This water-table takes different positions, as *bb*, depending upon the quantity of surplus water in the soil. In this kind of soil the drains may be placed 100 feet apart. Suppose that the drains are placed at this distance in a very retentive soil, as is one which is largely composed of clay, and we have left between the drains a portion

of soil which is undrained. This is shown by the line
cc. The water-table rises from the drain more ab-
ruptly, owing to the greater resistance of the soil,
which limits the width of land acted upon by the
drain. This requires the drains to be placed nearer
together. The line *cc* rises higher between the drains,
because of the nature of the soil; hence the drains
must be placed nearer together. It is not necessary,
perhaps, to give any demonstration of the curves from
laws which govern them. The statement of the fact
of their existence is all we shall attempt at present.

We have been more lengthy in the discussion of
principles than in laying down definite rules, for the
reason that a good general knowledge of the principles will enable the farmer to determine the matter for
his own particular case, where definite rules would not
apply.

SIZE OF TILES.

It has been previously stated that the grade of the
drain affects the velocity of the flow of water, and consequently the quantity which will be discharged in a
given time. In consequence of this the grades of the
drains which we wish to lay should be known before
we can determine the most economical size of the tile
to be used in order to drain a field or farm. In the
consideration of this subject we shall endeavor to give
a few practical directions, which have been found to be
reliable, without entering into the mathematical demonstration of formulæ, which would not be of use to
the farmer in ordinary drainage. The questions to be
taken into consideration relating to this subject are:

1. What is the area to be drained?

2. What is the greatest rainfall upon that area in
twenty-four hours?

3. What is the amount of surplus water which must be removed from the soil by drains, compared with the rainfall?

In the case of *casual* or *random* drainage, by which we mean the laying of a line here and there to drain some sag or wet place, and of which there is a great deal done, and often necessarily, there is a much greater area to be drained by one line of tile than we are apt to suppose. The one line will act directly upon a strip of land fifty feet wide on either side of it (in prairie soil), giving thorough drainage, and indirectly on all the land whose surface slopes toward the drain. In case of ponds and sloughs, a great quantity of water, in times of heavy rains, passes very rapidly over the surface of surrounding slopes and gathers upon the lowest land through which the drain passes. There is also a constant percolation of water through the soil upon the slope towards the soil which is directly acted upon by the drain. This has before been described as natural under-drainage. This being the case, we have found by experience that the area which we should consider in fixing upon the size of the tile is, in addition to the land acted upon directly by each drain, about one-third of all the land beyond this which slopes toward the drain, provided the slope is three feet or more in one hundred feet. The less the slope the more we may decrease this amount. For example, suppose the land, which is directly drained (taking fifty feet each side of all the drains), is two acres, and the sloping area beyond this is nine acres, then the area for which we must provide drainage is about five acres. A failure to consider the drainage area in this way has often led to the use of tiles which are too small.

If the land is flat and the drains laid out systematically, the area drained is easily determined. In such cases we need only to remove the water which falls upon the district itself. If, however, this district has land around it which slopes toward it, thereby throwing water upon it which does not properly belong there, we must regard the area in the same way as in the case of casual drainage. The *average* rainfall does not enter into this computation, but the greatest rainfall at any one time. If we can provide for the removal of the surplus water which falls during twenty-four hours, in the succeeding twenty-four hours, it will not do serious injury to the crops.

To find what this maximum quantity is, an actual record of the rainfall at this place (Tonica, La Salle county, Ill.) during the summer of 1880, will serve to show what we may expect. The table gives the rainfall in inches for the twenty-four hours previous to the morning of the day given in the column of dates :

APRIL.		MAY.		JUNE.		JULY.		AUGUST.	
Date.	Inches.	Date.	Inches	Date.	Inches.	Date.	Inches.	Date	Inches.
Apr. 2	0.08	May 9	1.20	Jun. 6	0.80	July 5	0.90	Aug. 2	0.30
" 3	0.23	" 10	0.94	" 8	0.10	" 8	0.30	" 20	0.28
" 4	0.21	" 20	0.28	" 9	0.37	" 19	0.20	" 24	0.63
" 14	0.09	" 21	0.09	" 14	1.45	" 20	0.02	" 25	0.05
" 16	0.45	" 27	0.00	" 15	0.40	" 23	0.44	" 27	0.65
" 17	0.28	" 30	1.79	" 25	0.64	Aug. 1	0.10	" 28	2.10
" 19	0.36	" 31	1.18	" 27	0.19			" 29	1.80
" 24	0.70	Jun. 1	0 30	" 30	0.50			" 30	0.30
" 25	0.56			July 1	0.14			Sept. 1	0.20
" 26	0.02								
" 28	0.18								

It will be seen that there are five days in which the rainfall was over an inch, and one day when it reached 2.1 inches. It is usually considered that one inch of rain in twenty-four hours is the maximum for which it is necessary to provide. The very excessive rains are quite apt to come after the soil is quite dry, and a great amount is absorbed. As this is not always the case, we should not be safe unless we assumed that one and a half inches would at times fall during twenty-four hours. This will give us 40,731 gallons which fall upon one acre of land.

The next question is, what part of this water is used by plants and carried off by evaporation, and what part must be removed by drainage? Many experiments have been made to determine this, and the amount discharged by drains has been found to vary much with the soil. We may say, in general terms, that about half the rainfall should be carried off through the drains. For a rainfall of one and a half inches we must remove 20,365 gallons of water from each acre, and this must all pass through at least a part of the main drain. The depth to which the land is drained and the nature of the soil will vary the conditions, so that the amount of water to be taken off may be much less. The fact that the soil when drained to a depth of three or four feet will hold an immense quantity of water, which will not, for the time, interfere with growing crops, allows us to use much smaller tiles than if we were required to remove all of the surplus water in twenty-four hours, and also renders close calculation as to size very difficult. As noted before, deep drainage requires tiles of less capacity for the same area than shallow drainage. The following directions may be given as a general guide in regard

to the size of the main to be used for a given number of
acres. We will take as a basis drains laid not less than
three feet deep and on a grade of not less than three
inches in one hundred feet. For drains not more than
five hundred feet long a two-inch tile will drain two
acres. Lines more than five hundred feet long should
not be laid of two-inch tile. A three-inch tile will
drain five acres, and should not be, of greater length
than one thousand feet.

A four-inch tile will drain twelve acres.

A five-inch " " " twenty acres.

A six-inch " " " forty acres.

A seven-inch " " " sixty acres.

A long drain has a less carrying capacity than a
short drain of the same size tile laid upon the same
grade. In order that the long one shall do as effective
work as the short one, larger tiles must be used if the
grade remains the same. If we double the grade per
one hundred feet of the drain we increase its carrying
capacity about one-third. Hence, the steeper the grade
the smaller the tile required to do the same work. In
the above we have had reference to the size of the
main. The size of sub-mains and branches must be
proportionate to the size of the main, taking into con- .
sideration the fact that the capacities of tiles laid upon
the same grade are to each other as the squares of their
diameters. Thus the capacity of a two-inch tile is to
the capacity of a four-inch tile as four to sixteen. The
size of the tile should diminish toward the upper end
of the main or branch according to the decrease in the
amount of water which will need to pass through that
portion of it. In case the drains are laid only from
twenty inches to thirty inches deep, we can not expect
that the sizes we have given will do satisfactory work,

especially in times of heavy rain. The instructions given upon this subject are as definite as is desirable to give, unless we take up special cases, in which we must vary the general rules. A man who has had large experience in laying out drains would, of course, do it more economically than the inexperienced, for he could take into account the grades upon which the drains were to be laid, the nature of the soil, and the area to be drained. In this subject we have not in all cases given reasons for statements made, since that would take more space than seems desirable at this time, but they may be relied upon as generally true respecting the subject under consideration.

KIND OF TILE.

The tile selected should be well burned, being hard enough to ring when struck with a knife blade. It is not well to get those which have been drawn out of shape by excessive heat in burning. They should be smooth on the inside, as the friction will be less. The best shaped tile, all things considered, is that in which the cross section is a circle. They can be laid more easily and give greater capacity for the material used. The requsites then are circular tile, of good clay, well burned, smooth and true in shape.

CONCRETE TILE.

While tiles made of burned clay have been well tried and are safe where draining is practicable, a concrete tile has been introduced during the past year, which promises to serve an excellent purpose. It is manufactured by means of a simple machine, which is operated in the ditch after it has been dug and graded in the usual way. The materials used are the best quality of hydraulic cement, lime, and coarse sand.

These are mixed in proportions suitable to the work and made into a stiff mortar. The mortar is fed into the machine through a hopper and comes out at the rear end of the machine a continuous pipe, smooth inside and outside. By means of a trowel made for the purpose, the continuous pipe thus made is cut into sections of any desired length, in such a way that the bottom is left continuous, yet sufficient crevices are left for the entrance of water. The cement soon sets, and in a day is hard enough to bear the weight of the filling without crushing. When fully hardened, the tile has the durability of stone. The machine and material are not very expensive, and the farmer will find it to his advantage to investigate this new tile and its manufacture. So far, it lacks the tests of time and experience to prove its desirability, yet in the writer's opinion it will serve an excellent purpose in many places, especially for continuous water pipes about the barn and dwelling.

LIBRARY OF THE UNIVERSITY OF CALIFORNIA.

CHAPTER V.

PRACTICAL DETAILS OF THE WORK.

Mapping Drains—Construction of Drains, Grading the Bottom, Outlet, Laying Tile—Difficulties iu Constructing Drains—Obstructions to Drains—Junctions.

MAPPING THE DRAINS.

The drains having been staked out upon the ground, the grades arranged, and the size and number of tile fixed upon, we should make a map of the drains which will show their position, length, fall per 100 feet, and the physical features of the land through which they pass. This, with the notes, will give all the information respecting the drains which it will be necessary to preserve. The map is merely a sketch showing the position and length of the drains, like the one shown in fig. 16; can be easily made; will show what has been done, and will serve as a record of the improvements in the drainage line, just as others are shown by houses, barns and fences. One may think he does not care for a map of his drains, as they will show themselves in the condition and improvement of the soil, yet when he begins to forget their location he will wish that he had some representation of them to refresh his memory and to show his friends, if nothing more.

THE CONSTRUCTION OF DRAINS.

The work done thus far has been preparatory to the actual digging of the ditch and laying of the tile. It will seem to many that the staking out, leveling and adjusting the grade of drains, is too much work for little pay, if not wholly useless. The farmer, perhaps, has seen his neighbor do some draining "by guess" which has worked well, or he may have laid a short line of tile himself with good results, by simply using the running water as a guide. In more extended sys-

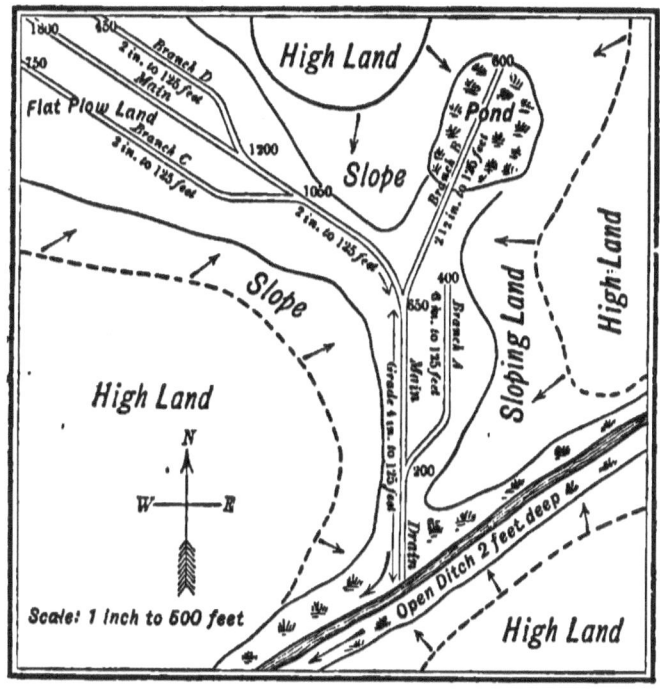

Fig. 16.—Map of a Drained Field.

tems, and with no water, or even with water for a guide, he will sometimes partially or entirely fail. "Be sure you're right and then go ahead," is the motto to use in draining. With the drains laid out and depths marked upon the stakes, and every thing arranged as it should be, we know in advance that the drains will work perfectly, if they are laid according to the survey. If the whole system is not completed at one

time, a part can be done in one year and the rest the next, or any other convenient time, though it is better to do it all at once, as the frost will move the grade pegs so much that the unfinished part will be obliged
. to be leveled again. It is better, however, to do part of the leveling twice if necessary, rather than not have the drains laid out upon some connected system.

Digging the Ditch.—The tools which are necessary to do good and rapid work are, First, a ditching spade for the main part of the digging. This spade has a

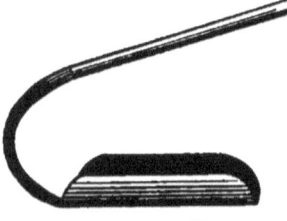

blade about eighteen inches long, a little narrower than the common spade, and curved towards the front. The superiority of this over the common spade is that it is capable of taking narrower and deeper

Fig. 17 - Tile Hoe

drafts, which will adhere to the spade until it can be lifted out. Secondly, a tile spade, which is narrower than the ditching spade, and tapers towards the point. Third, a "pull scoop," or tile-hoe, for cleaning the bottom of the ditch. The handle of this should be at such an angle that the workman can use it when standing in the ditch sixteen inches from the bottom. As this is a very handy tool, and it is often necessary to get it made by the blacksmith, we give a drawing of it (fig. 17). It is most convenient when it is just large enough to make a channel into which the tile to be used will fit firmly. This would necessitate a different hoe for every different sized tile. A tile-hoe suited to three-inch tiles may be used for much larger sizes, with a little care and extra work. Fourth, a gauge, which is a stick six feet long, and one and one-fourth inches square in section, graduated to feet, inches and eighths of inches, and having an arm two feet long, which slides up and down upon the graduated stick. The arm should move at a right angle to the stick and fasten at any desired point by means of a thumb-screw. Fifth, a strong hemp line 100 feet long, for lining out the ditch, and also for a gauge line to be used in grading the bottom of the ditch.

Begin opening the ditch at the outlet. Stretch the line about four inches from the grade pegs, and take out one draft with the ditching spade, making the ditch ten inches to twelve inches wide, where the ditch does not exceed four feet in depth. The ditch should be clean cut and straight on the sides. If there are short crooks when it is started at the surface, they are apt to increase as the ditch is deepened, so that the final channel for the tile becomes crooked to a troublesome degree. The workman should have in mind the depth which is marked upon the stakes, between which he is working, and in digging he should aim to leave a bottom spading of sixteen or eighteen inches for the one who finishes and grades the bottom. We are supposing that the soil works easily, not being hard enough to require picking, when we give ten or twelve inches for the width of the ditch at the top. If the soil is so hard that the spade can not be pushed down its full length, or it must be loosened with a pick, more width must be allowed—sometimes even twenty inches—at the top, the ditch in all cases slanting in, of course, toward the bottom, to the width of the tile.

Grading the Bottom.—The ditch having been dug to within sixteen inches of the bottom, as near as the workman can judge, we must take out the last draft with the tile spade, and bring the bottom to a true grade and depth, as indicated by the stakes. These, we remember, give the depth measured from the grade-peg. It remains for us to connect these points so that the bottom of the ditch shall be a true line upon which we may lay the tile. There are many ways of doing this

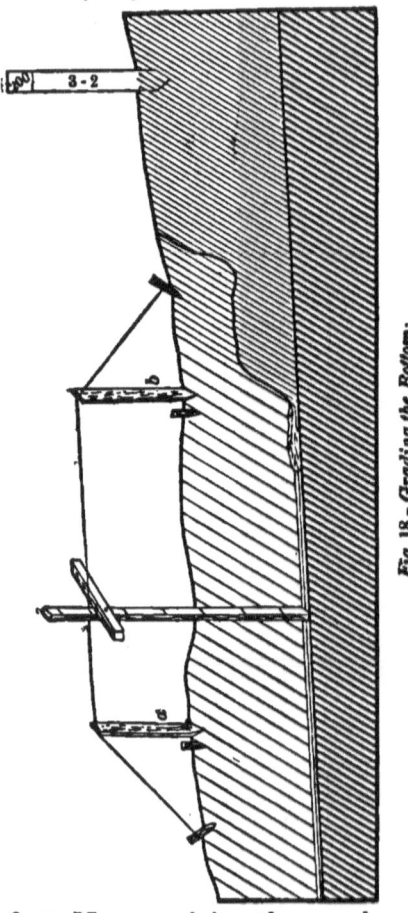

Fig. 18 - Grading the Bottom.

correctly, but we give only one, which has been found simple, easy, practical and correct, and which is a favorite method with all who have used it. It consists in stretching a line at the side of the ditch at any convenient height parallel to the required bottom of the ditch, as shown in fig. 18. In order to do this, we take the gauge before described, a convenient length of which is six feet, though, of course, for very deep ditches it must be longer. In a four-foot ditch we may set the arm of the gauge at six feet. Now, noticing the number of feet and inches marked upon the first stake, subtract it from six feet, and note the difference. Drive a stake, a, by the side of the grade-peg until the distance from the top of the peg to the top of the stake a is the same as this difference. Set another stake at the next grade-peg in the same way. Stretch the line over the tops of these stakes and fasten by small pegs, as shown in the cut. The line is now parallel to the line upon which we wish to lay the tile, and six feet above it. It may be set any other convenient distance above the bottom by the same method. If the line sags, place a stake under its middle. Take out the last draft with the tile-spade, being careful not to go too deep. Dig as far at a time only as you can conveniently reach back over with the tile-hoe, and then, without stepping down to the bottom of the ditch, clean out the loose dirt with the tile-hoe. Have the gauge at hand, and holding it in a vertical position, see what is necessary to make the bottom parallel with the line. Then pare the bottom down until the arm of the gauge when again placed will just graze the line. Test every foot along the bottom, and do not leave it until it is right. A little water in the bottom of the ditch causes the soil to work easier, but more than two inches is a hindrance to good work. The digging may go on in this way, the workman being careful not to walk in the ditch after it is prepared for the tile. The line is set from stake to stake as the work proceeds, due care being taken that everything is right.

Laying the Tile.—Before the ditch has been graded far, it is best to lay and cover the tiles, to prevent hindrance by the falling in of earth. Begin at the outlet and lay the tiles by hand, turning them about until the ends fit closely upon the top. If the bottom is firm and hard, the workman may stand upon the part already laid, for the tiles should fit true and solid in the groove made for them. Be sure that they are turned until they fit well. This is an important part of the work, and must not be slighted. When the work of laying stops, while more ditch is being prepared, the

5

upper end of the drain should be protected from mud which may be washed down as the digging proceeds. If the bottom of the ditch is soft, the tiles may be laid from the surface with a tile-hook. Workmen who do job work prefer this, as the work is done easier and faster, but for excellence of work we prefer the hand laying. The upper end of each drain should be closed by placing a stone or brick over the end of the tile before covering. The tiles when laid should be imme-

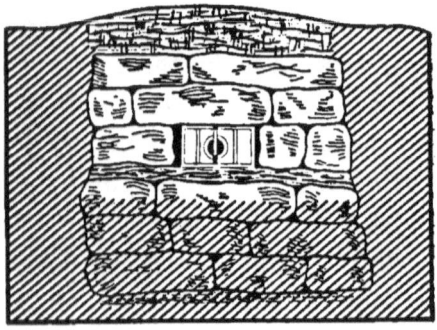

Fig. 19 - Outlet

diately covered with moist clay, which has just been taken out, or is obtained by slicing off from the sides of the ditch with the spade. This should be firmly pressed about the tile with the feet to a depth of six inches.

One of the first things to attend to is to secure the outlet from injury by trampling of stock, etc., and the entrance of vermin. If long delayed, water may flow through the tile, and seriously interfere with any work at the outlet. We give an illustration of a good way to secure the outlet from injury in fig. 19. Dig a pit two feet square where the outlet is located, and lay a foundation of stone deep enough to be safe from frost. Build this up to the line of the drain, laying the stone in cement and sand mortar. Lay the first two tiles upon this, being careful that they connect correctly, and continue the stone work to within one foot of the surface. A grate may be placed a little distance in front of the end of the tile, and fastened as the masonry proceeds; or the outlet tile may .be larger than the others, and have wires passed through holes drilled in the tile for that purpose. Brick may be used instead of stone, or even wood will be good as long as it lasts.

After the tiles are first covered the ditch may be filled in in any way which may seem desirable. If the earth is dry, it can be done with about one-half the expense required when wet. All of the earth should be heaped upon the ditch, as when it settles there will not be too much. When done by hand a simple and expeditious method is to pass a rope around the blade of a large scoop-shovel, then with one man pulling at the rope on one side of the ditch, and another pushing upon the handle upon the other side, the earth can be moved rapidly.

We have described only one way of performing the work. Many practical workmen in drainage may take exceptions to it, and recommend a better way. Any method which is well understood and gives good results is apt to be regarded as the best. That which we have described has been demonstrated by actual use to be practical and correct.

DIFFICULTIES IN CONSTRUCTING DRAINS.

At some seasons ditches may be dug with less expense than at others; but the most favorable time can not always be selected. The clay subsoil may be dry and hard, necessitating slow work, and sometimes the

pick must be used. Under such difficulties the ditch
should be 14 to 20 inches wide at the top, that the
workman may have room for his arms and shoulders.
If, on the other hand, the earth is so wet that it will
"slump in," the sides must be sloped from the surface
at such an angle that they will stand. Another way to
overcome this difficulty in deep ditches is to "sheet,"
by driving two-inch planks endwise at the sides of the
ditch until the lower ends are below the bottom of the
required ditch. The planks may be kept in place by
cross braces above, and the earth taken out between the
lines of "sheeting," and the tile laid to grade. The
planks may be taken up and another section prepared
in the same way. This method is to be resorted to
only in deep ditches opened in very unstable and
treacherous soil.

If the soil or subsoil is a wet silt, quicksand, or
other material which "runs," the joints of the tiles
should be covered with firm clay, or a band of grass
or straw. This will keep such material out of the
drain until it has made the soil sufficiently dry and
firm to give no trouble from this source. If the drain
can be put in position and the earth compacted about
it without moving the tiles, there is little to fear concerning its success and durability.

It is often desirable to drain a spring, but the earth
is usually so soft that the tiles can not be placed in
position. In such cases a bottom can be most easily
made by placing a fence-board for the tiles to rest
upon, observing to have the tiles laid upon the proper
grade, and to cover them with a few inches of firm
earth to hold them in place and prevent mud from
obstructing the drain. Gravel-stone or clay may be

used for a bottom with good results, but they are not always easily obtained.

OBSTRUCTIONS TO DRAINS.

Drains are sometimes obstructed by the roots of such trees as the willow, the water elm, etc. Experience has fully proved that a tile drain is not safe near willows. The fine roots penetrate the line at the joints, and flourish with such luxuriance that in one or two years the tile will be entirely filled with rootlets. The only safe way is to destroy all such trees within seventy-five feet. For a properly constructed drain, the only care it requires is to keep the outlet free from mud, remove the silt which accumulates in the silt-basins, and to see that no trees with water-loving propensities are permitted to obstruct the drain with their roots.

JUNCTIONS.

Great care should be taken in making junctions with other drains, that the joints be good, and that the branch drains always discharge in the direction of the current of the main drain. Right-angled junction tiles should not be used. Junctions should be so made that the branch will enter at an angle of about 30° with the one into which it flows; or if a greater angle is necessary, the mouth of the branch tile should be curved. In fig. 20 are shown sections of junctions by which the truth of the above remarks may be made more apparent. No. 1 is a section of a right-angled junction. Suppose that the two currents meeting at *a* be represented by the lines *ab* and *ad;* the velocities being equal, the lines are made of equal length. Completing what is termed the parallelogram of forces, *abcd*, and drawing the resultant *ac*, we have the direc-

tion of the current resulting from· the union of the
two.   This resultant, it will be seen, flows strongly
against the opposite side of the tile, checking the current by the friction thus caused, and also creating a little eddy in which earthy material washed out by the
drain may be deposited.   If the velocity is greater in
the branch, as is often the case, we find the resultant
in the same way, by making the length of the lines
proportional to the velocity of the currents we have to
represent.   Supposing the velocity of the current in

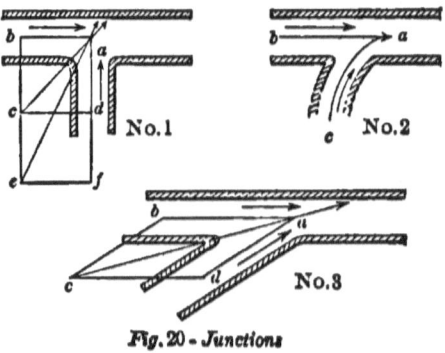

Fig. 20 - Junctions

the branch to be twice that in the main.   The resultant *ae*, in No. 1, shows that there is still greater resistance offered than in the first case assumed.   The best
form for a junction which joins a branch to a main at
nearly right angles, is shown in No. 2.   This curves
the current gradually as it enters the main stream,
uniting it with the main current in a way that acceler-
ates it rather than retards it at the time the two unite.
No. 3 shows the resultant of the two currents when
they unite at an angle of 30°, which is the preferable
angle.   A study of these figures will show the necessity of careful attention to this subject.

Many mistakes have been, and are still being made

by western farmers in the practice of drainage. Among them we notice, first, faulty and insufficient outlets. This is a difficulty always present where the land is flat. The remedy is for farmers to unite and construct large and deep open ditches, into which tile-drains may discharge freely. Second, the use of tiles too small for their subsequent work. The sizes given in a previous chapter are small enough. If the grade upon which they are to be laid is decreased, the size of the tiles must be increased, in order to do the same service.

We have considered the subject of drainage simply in its application to land for agricultural purposes. We have not attempted to exhaust the subject, or to discuss a variety of opinions and practices, but to give simply the principles underlying it, and a practical method of carrying them out. The methods of calculation used by drainage engineers have not been entered into, only to an extent which may be of use to the careful farmer. The cost of drainage and the profit to be derived from it are matters upon which almost every one can satisfy himself by consulting with those in his vicinity who have had some experience in the work. If whatever is done be done well, we have no fear in asserting that the result will surprise our western farmers, and will so add value to prairie farms for agricultural purposes that it will be considered the most important improvement that can be made upon them.

CHAPTER VI.

DITCHING MACHINES.

Difficulties Involved—Principles—The Johnson Tile Ditcher—
The Blickensderfer Tile Drain Ditcher.

There has been considerable effort, during the last
few years, to lessen the expense of ditching for tile-
drains by the invention and use of machines. The
success thus far attained is very encouraging, and it is
to be hoped that soon the machine will very largely
supplement hand-work in ditching. There are many difficulties incident to the digging of farm drains. The
soil is often soft and sticky; at other times it is hard,
and in some localities contains gravel and stone; deep
cuts must sometimes be made; springs must be drained;
in short, the difficulties to be overcome by the inventor
of a tile ditching machine can hardly be appreciated
by any one unacquainted with practical ditching.

There are essentially two different principles upon
which the machine problem is being worked out. One
is the repeating process, by which the machine is made
to pass back and forth over the ditch, each time adding to its depth until it is completed. The other com-
pletes the ditch as the machine advances, requiring
only one passage over the ground.

The following cut represents a repeating machine,
called the Johnson Tile Ditcher, manufactured by King,
Hamilton & Co., Ottawa, Illinois. The machine is
drawn by eight horses, four abreast. A revolving
wheel containing small spades loosens the earth and it

is elevated and thrown on one side of the ditch. At each passage of the machine over the ground, the depth of cut is uniform and the ditch is completed with a

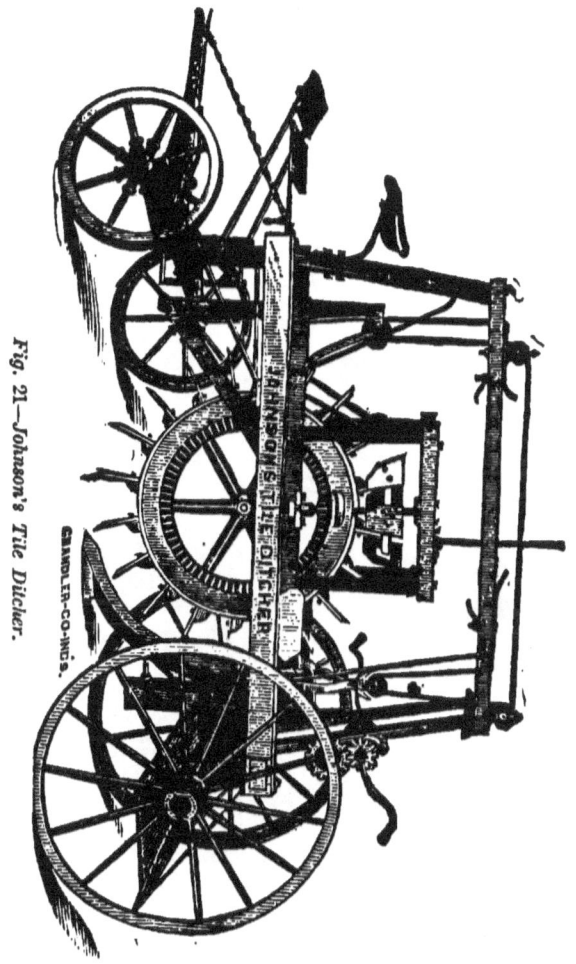

Fig. 21—Johnson's Tile Ditcher.

regular grade where the necessary care is observed. The manufacturers claim that they can cut from one hundred to one hundred and fifty rods of three-foot ditch per day with this machine, and leave the bottom in good shape to receive the tile.

Fig. 22.—The Blickensderfer Tile Drain Ditching Machine.

The cut on page 74, represents the Blickensderfer Tile Drain Ditching Machine, manufactured by U. Blickensderfer, Decatur, Illinois,* which consists of a large revolving wheel of excavators or buckets mounted upon four wheels. The buckets are of a new and peculiar shape of construction; they grow larger inside from their mouth or cutting edges back, and as the clay is scooped up and passed in, not being compressed, but getting into larger space in the back of the buckets, it readily drops out when revolved to the top of the wheel, the earth being thrown on an inclined apron to one side of the ditch. Curved teeth or picks of steel, projecting beyond the buckets, loosen hard-pan clays and protect the buckets from stones, which they · also lift and work out by a simple manipulation of the revolving wheel, the forward motion of the machine being readily checked or backed during the operation without stopping the horse.

The manufacturer of this machine claims that it will cut a ditch over four feet deep in one passage over the ground, and at the same time give any desired grade over an uneven surface, the grade being as perfect as a person can sight the top of a row of stakes set to indicate the desired grade.

The power consists of a single horse, used upon a sweep around the machine, which revolves the buckets and at the same time moves the machine ahead, requiring only one man and a boy to attend it. It will cut from seventy-five to one hundred rods of ditch per day, finishing it accurately for the tile. The bottom of the ditch is shaped so that a large or small tile will fit it and not roll or get misplaced.

These, as well as other machines for digging ditches,

*Formerly Erie, Pa.

are worthy of close attention on the part of tile drain-
ers. If they are not yet already satisfactory in every
respect, it is quite probable that one or all of them
will soon take the place of hand-ditching, at least in
many kinds of soil. , The outlook for reclaiming our
waste lands and the improvement of cultivated farms
is very promising when we consider the impetus which
successful tile and ditch machinery is giving to this
progressive work. The old adage "Nothing succeeds
like success" has proven true in the history of draining, and will also be true of the ditching machine
which once shows to the people that it is a success in
the way of doing good work at less cost than it can
· be done by ordinary hand work.

CHAPTER VII.

COST AND PROFIT.

Cost of Drainage—Cost of Mains—Profits of Drainage.

COST OF DRAINING.

The first cost of draining is what frightens many farmers when the subject is brought to their notice. Draining should be regarded as an investment of capital. The farmer's land, his necessary stock and implements, and his yearly labor are regarded as his capital. All that he can make by the management of these is the profit of his business. But to drain, cash capital is required. If the farmer does not possess this, and can not get it at a reasonable rate of interest, he can not drain. If he has been prosperous, and, as a result, he has in his possession a little cash capital, he does not hesitate to use it in adding to his facilities for increasing his profits. Nor does he hesitate long to pay a reasonable rate of interest for money with which to add to his working force if he can see that there is a fair prospect of making a much larger profit than the rate of interest he is obliged to pay.

Without naming a definite number of dollars and cents, let us take this general statement, which is admitted by all who have drained to any extent, that on ordinary farm land that will produce but one-fourth to one-half a crop, the total cost of draining will be met by the additional crop that will be produced during the next two years after draining. The cost of cultivating drained land is less than for wet land, as

all who have tried both will admit. By this investment the farmer will get fair wages for what work he
does upon his land, his money will be in his pocket
again at the end of two years, and his land drained
and ready to return his money every succeeding two
years. Many farmers cultivate land which, in reality,
does not pay them a fair remuneration in the crop
they get from it. Their profits come from land which
is in good condition and will produce good crops.
Very often they would make more to wholly discard
the wet land and give more labor to that which will
give some return for it.

Where land is worth fifty dollars per acre, it will
pay a large return to drain wet land. It is true that
much western farming is done on cheap land where it
is the custom to cultivate a part of the farm, which is
naturally surface-drained, and use the rest for grazing
purposes. Draining under such conditions will not
pay, because the farmer has not the facilities for using
the good land he already has in his possession. Such
farming, however, is fast coming under the more ad-
vanced system in which "more work and less land" is
the motto.

The cost of draining, like any other enterprise undertaken upon the farm, will always vary with the price
of labor, yet farm products usually bear a price
commensurate with farm labor, so that the relation
of the two will be about the same. There are two
things which will vary the expense of draining whatever may be the relation between farm products and
farm labor. These are: the main drains, which will
necessarily vary in size, length and depth, and the degree of thoroughness with which draining is done.

COST OF MAINS.

It will be easily understood that a field or farm may be so situated that very little expense will be required for large mains into which to discharge the laterals. It may be for the reason that the field is near some large open ditch, or some stream which is easily reached. Draining will then be reduced to a minimum expense. There are often cases where a main drain of considerable size must be long and laid deep in order to give the necessary outlet to the field.

It is the custom of ditchers to dig ditches and lay tile by the rod, though a more convenient unit would be the foot or one hundred feet. As a basis, we will say that ordinary diggers can be had for $1.50 per day, and good ditchers at $2.00 per day. For a main drain, laid at different depths and with tile of different sizes, the expense per one hundred feet is as follows:

Cost of Five-inch Main per One Hundred Feet.

Depth of Ditch.	Cost of Digging and Laying.	Cost of Tile.	Cost of Filling Ditch.	Total Cost per 100 feet.
3 feet.	$1 50	$3 00	30 cents.	$4 80
4 feet.	2 00	3 00	42 cents.	5 42
5 feet.	3 00	3 00	60 cents.	6 60
6 feet.	4 50	3 00	75 cents.	8 25

Cost of Six-inch Main per One Hundred Feet.

Depth of Ditch.	Cost of Digging and Laying.	Cost of Tile.	Cost of Filling Ditch.	Total Cost per 100 feet.
3 feet.	$1 50	$4 00	30 cents.	$5 80
4 feet.	2 10	4 00	42 cents.	6 52
5 feet.	3 00	4 00	66 cents.	7 66
6 feet.	5 10	4 00	78 cents.	9 88

Cost of Seven-inch Main per One Hundred Feet.

Depth of Ditch.	Cost of Digging and Laying.	Cost of Tile.	Cost of Filling Ditch.	Total Cost per 100 feet.
3 feet.	$1 80	$6 00	36 cents.	$8 16
4 feet.	2 40	6 00	48 cents.	8 88
5 feet.	3 00	6 00	72 cents.	9 72
6 feet.	5 70	6 00	90 cents.	12 60

Cost of Eight-inch Main per One Hundred Feet.

Depth of Ditch.	Cost of Digging and Laying.	Cost of Tile.	Cost of Filling Ditch.	Total Cost per 100 feet.
3 feet.	$1 92	$8 50	42 cents.	$10 84
4 feet.	2 58	8 50	54 cents.	11 62
5 feet.	3 90	8 50	78 cents.	13 18
6 feet.	6 00	8 50	$1 00	15 52

These tables give a pretty close estimate of the cost of mains in general, when wages are two dollars per day for good ditchers. To this should be added the cost of boarding the men while they are at work, and of hauling the tile from the factory, or station to which they are shipped. The estimate for filling the ditches is made on the basis of part being done by hand and part by team work. It is often the case that ditches can be filled with but little expense by using a steady team with a plow and scraper. When it is necessary to dig the ditch five or six feet deep, the risk of striking rocks, hard clay or quicksand makes the estimate more unreliable, and it will be found in such cases that the cost of digging the ditch and laying the tile will run over, rather than under, the estimate.

It should be clearly understood that these items of cost will vary greatly with different years and in different localities. The cost of tile and labor are not the same for two years in succession, hence the foregoing estimate must be varied with such changes.

The actual cost to the farmer may be greatly diminished by "taking time by the forelock," getting his plans well laid, and when his farm help is not rushed with work, let them devote their time to the drains. There are winters which are often so free from cold weather that ditching can be done to good advantage if the tile have been previously hauled. Help is then plenty and cheaper than during the summer months. There are other seasons of the year when the forces of the farm can be used with economy in carrying on drainage work.

COST OF BRANCH DRAINS.

Branch drains, laid from three feet to three and a half feet deep, in ordinary farm land, which will spade easily, will cost $2.00 per one hundred feet for digging the ditch, laying the tile and filling up

6

the ditch. Three-inch and four-inch tile cost from
$1.32 to $2.00 per one hundred feet respectively.
Add to this the cost of boarding the men while en-
gaged in the work, and of hauling the tile to the
ground, and we have a close approximation of the
cost of draining per one hundred feet. There are a
few incidental matters, such as protecting the outlet
tile, silt basins, if any are needed, and surveying, which
should be taken into account.

The actual cost per acre will depend upon how many
rods of drain are laid upon an acre. A field having
several wet places and always troublesome to cultivate
in the spring, to say nothing of the loss incurred, can
often be drained out in good shape at a cost of about
five dollars per acre for the whole field. The cost of
draining a field by placing the laterals from sixty feet
to one hundred feet apart, making a fair allowance for
the extra cost of mains, is from fifteen to twenty dollars per acre, at prices as they exist at the present writ-
ing (1882).

PROFITS OF DRAINING.

Enough has been said incidentally, in the foregoing
pages, concerning the profits of draining, to satisfy
owners of wet farming lands, that it is better to re-
claim such lands than to invest capital by buying new
farms. After having arrived at the probable cost of
the work, it will be easy to estimate the increase of
crops by comparing the field to be drained with one
which is naturally drained, and whose productive pow-
ers have been well ascertained. If the land is in such
a condition that it will produce nothing without draining, the entire crop, after deducting the cost of producing, will be profit. We will suppose that an undrained field will produce twenty bushels of corn per

acre. If well drained, the same labor and expense in cultivating will produce fifty bushels of corn per acre. Here is a gain of thirty bushels per acre, which, at forty cents per bushel, will be twelve dollars per acre as profit. The labor, which in the first case produces eight dollars, in the second brings twenty dollars. Other crops can easily be figured in the same way, and such results are realized every year by farmers who have drained in an economical and thorough way. The farmer will have more confidence in his work if he investigates and figures for himself. Let him estimate the cost carefully and compare it with the expected gain. We do not advise any one to go at this work blindly, or because others say it is all right. Count the cost and the expected increase fairly, and then shape your course accordingly.

We can give in a sentence the condensed evidence of scores of farmers upon this subject of profit, viz: that *draining pays from twenty-five per cent. to fifty per cent. on the investment.* The writer regards it as almost superfluous to publish individual statements regarding this subject, as they can be found in almost every paper published in the interest of farmers. The aim of the author has been to tell how to drain in such a way that the best results may be obtained. The profits will be assured if the work is adapted to the case in hand and well done.

CHAPTER VIII.

ROAD DRAINAGE.

Improvement of Roads—Surface Drainage—Under-Drainage—
Effect of Tile Drains upon Roads—Care of Drained Roads.

ROAD DRAINAGE.

While the drainage of farm lands is of great importance as an aid in the production of large crops, the drainage and other improvement of our public roads may be regarded as meriting equal attention, since the farmer must have an opportunity to market his products, which is commensurate with their quantity.

Western roads are often impassable for loads during several months of the year by reason of the mud. There are, of course, parts of the West in which the roads are graveled, or are by nature good nearly all the year. To improve our roads, as we have them now, is an important problem. The arguments for gravel roads on prairie land have very little force considering the great distance from which gravel and stone must, in most cases, be brought, and the consequent expense. Were the gravel at hand, it could not be used for the construction of roads until a firm foundation had been prepared for it by drainage.

Let us take the roads as they are and improve them by using the means within our reach. All that has usually been done to our roads in the way of improvement has been to make sloughs, ponds and swamps passable. An embankment has been made through these places by scraping the earth from either side towards the middle of the road, leaving ditches about two feet deep, and making an embankment two or three feet high. Plank culverts are used in the sloughs to allow the water to pass through the embankment. The side ditches are often made with little reference to the discharge of the water which they collect from the road and adjacent fields. As a consequence these ditches remain full of water during a large part of the spring, saturating the bottom or the embankment until it is in a soft and plastic condition. The surface of the road becomes broken up by the frost as it comes out of the ground, the spring rains penetrate it until the saturated surface meets that of the bottom, and we have a mass of mud impassable as a road, and a discouraging contrast to the smooth and firm road enjoyed by travelers the summer before. As the water evaporates, or slowly drains off through the soil, the road begins to dry and in process of time becomes firm but rough, the track full of ruts, which will hold water to their full capacity at every rain-fall, while the embankment is flattened, and perhaps hollow in the middle. The road is annually repaired by raising the embankment a little with newly dug earth, and the process goes on from year to year.

IMPROVEMENT OF ROADS.

As before stated, we wish to take our roads as they are, and make them better by using the means within our reach. We may have our ideas of what perfect roads ought to be, and show what has been done towards their attainment in other localities; but yet our own roads will remain as bad as ever until we hit upon some system which will apply to our soil and particular need.

Owing to the readiness with which our soil absorbs water, there will be a time in the spring of the year when the frost is coming out of the ground, that the surface will be soft and not fit to travel upon, do what we will; but the length of the time during which this is the case may be greatly shortened by a proper system of drainage. As roads are now worked, they are unfit for heavy travel for two months or more in the spring. If drained, this time can be decreased at least one-half, the road be much better for the remainder of the year, and the expense for repairs be reduced to a minimum.

SURFACE DRAINAGE.

The most essential feature of a good road on prairie soil is a hard and smooth surface. When a road track upon prairie loam is examined, it will be found that for a depth of about ten inches there is a hard and compact crust, while below, the soil is as loose as that in a meadow. The value of the road depends upon keeping this crust intact. The road bed should be raised and sloped sufficiently from the center line towards the ditches, to carry off the rain-fall. If the road bed is not deeply rutted, the rain will flow off so rapidly that the road crust will not be badly softened. The incline from the middle in an eighteen-foot road should be about six inches. If the road is not embanked, the storm-water, even on naturally drained roads, will run down the slope in the road track until it is spoiled, and a new one must be made. The side ditches are valuable in collecting the water of heavy rains, and suitable outlets should be provided for them.

UNDER-DRAINAGE.

The object of under-draining a road is simply to keep the bottom of the embankment firm at all times of the year. The open ditches should carry off the excess of storm water, but it remains for the tile-drains to remove what the ditches fail to carry off, and to give a dry road crust. In under-draining a road, first obtain a good, free outlet for whatever drains may be needed. If there is any doubt upon this point, use the level to find the fall above and below the proposed outlet, before any further arrangements are made. Do not make the outlet at some little ditch which will soon fill up, but be sure that all the water that will ever be discharged from the drains can flow away without "backing up." It may be that the outlet must be obtained through a farm adjoining the road. In such cases the owner of the farm

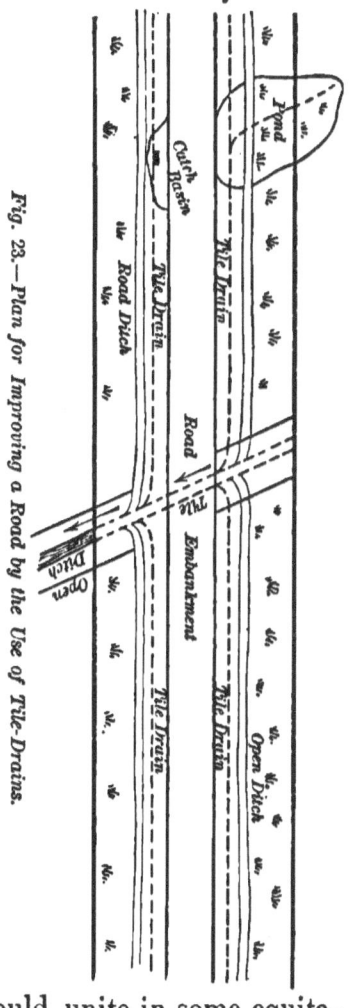

Fig. 23.—Plan for Improving a Road by the Use of Tile-Drains.

and the road authorities should unite in some equitable way.

Supposing that the road is in flat, swampy ground, lay out a drain lengthwise of the road on each side of the embankment and close to its base, as shown in the

plan (fig. 23) and cross-section (fig. 24). The reasons
for laying a drain at each side of the embankment, instead of one in the middle or on one side, are as follows: If one drain is laid in the middle of the road,

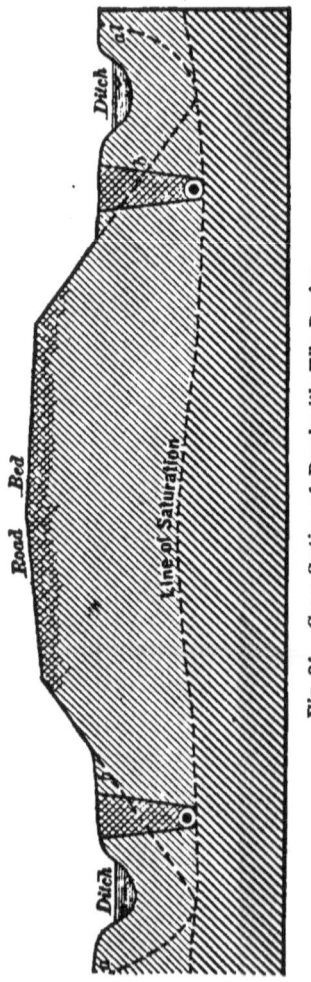

all water which enters it must
pass from the side ditches
through the base of the embankment. In times of high
water the drain will not
carry the water away quickly
enough, consequently the base
will be, for a time, saturated.
The idea that a drain in the
middle will remove the water
upon the road bed is errone-
ous, for at such times the mud
is puddled and will permit no
water to pass from the surface
to the drain. One drain at
the side of the embankment,
though excellent, is often insufficient to give good drainage. Two drains laid in the
way indicated in the figures
will prevent all water from
saturating the base of the embankment, even in the wettest
times. The drains should continue as far as the ground is
wet to any extent in the spring.

If there are ponds near by, which, in times of heavy
rains, overflow and discharge into the road, side drains
should be extended to them, as shown in figure 23.
Should there be a small hollow along the line of the

ditch, as sometimes happens, a catch-basin will facilitate the removal of the water. This is a pit two feet
square, dug as deep as the tiles are laid. After the
tiles are laid, the pit is filled with gravel and small
stones. The object of this basin is to take the water
which gathers so quickly in such places, and give it a
rapid ingress to the tile.

The depth of the drains should be as near three feet
as is consistent with the nature of the ground. As the
important thing to look after is the rapid removal of
water, special attention should be given to the manner
in which the work is done. The grades, if possible,
should not be less than three or four inches to one
hundred feet. The principles apply in this work as in
draining land for farm purposes, though the object
sought is different.

The size of the tile should be larger than would be
required for the draining of the soil for cultivation,
for the reason that storm water is thrown rapidly into
the side ditches from the surrounding soil, requiring
the tile to carry much more in a short time than if the
land were flat.

As suggestive of the size of the tile that it is best to
use, we may give the following directions: For drains
eight hundred to one thousand two hundred feet long,
use three-inch tile; one thousand two hundred to two
thousand feet long, four-inch tile; two thousand to
three thousand feet long, five-inch tile. These directions apply to the two lines of tile along the road, laid
upon a minimum grade of three inches per one hundred feet. If these do not discharge into an open
ditch, the size of the tile in the outlet drain should be
proportioned to the road drains.

EFFECT OF TILE DRAINS UPON THE ROAD.

The drains keep the embankment firm by preventing water from penetrating it from the bottom and
sides. The whole of the road is kept dry, except the
crust at the top, which,
if traveled upon when
wet, becomes puddled
and will allow no water
to go through. This
part of the road bed
must be made dry by
surface drainage and
evaporation. During
the fall and winter the
whole road embank-
ment becomes thor-
oughly drained, so that
there is only a little
frost to come out and
soften the surface of the
road in the spring. If
we could make em-
bankments along our
roads four or five feet
high, we should con-
sider our roads well
drained. Instead of
this, we withdraw the
water to a depth of four

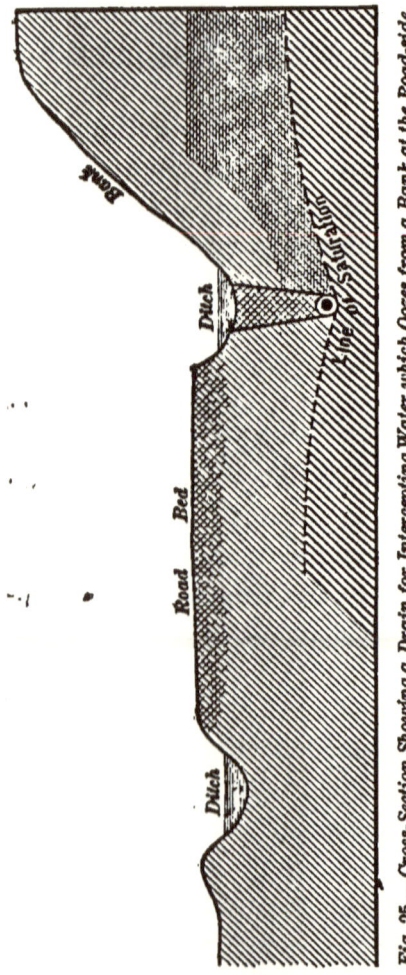

Fig. 25.—Cross-Section Showing a Drain for Intercepting Water which Oozes from a Bank at the Road-side.

or five feet from the surface of the road, which gives
it the effect of an embankment, as shown by the dot-
ted lines *a b* in fig. 24. Draining does even better
than this, for in the case of the raised embankment
water often stands in the large ditches at the side; but

when the under-drains are laid, no water at any time stands higher than the drains.

It should be borne in mind that under-draining does not prevent rain from penetrating the surface and allowing the road to be cut into ruts; but this affects only the surface and for a comparatively short time.

A tile-drain may be used in many places to intercept water which percolates through banks lying higher than the roadway. One line of tile, laid between the bank and road, as shown in fig. 25, will often cut off the water at such a depth as to render a road firm which has previously been soft. Springs of water often make muddy spots in a road. The sources of such water should be hunted up, and drains laid where they will convey the water from the road.

The only hope we have for good roads on our prairie soil is to keep them dry by removing the water as quickly as possible from the surface, and preventing the substratum from becoming saturated.

CARE OF A DRAINED ROAD.

After the road bed has once been put into proper shape, and well drained, nothing should be done except to keep the surface in shape and as smooth as possible. The old crust is better than any new, and should be preserved with the utmost care. All improvement of the surface should be upon the "stitch in time" principle. When the road becomes rutted out and begins to dry, it should be smoothed by drawing over it some machine made for the purpose, of which there are several. If the drains are properly constructed, the outlet attended to, as in the instructions given on farm drainage, the road will require but a small outlay for improvements each year.

Road drainage has been experimented upon until its benefits have been fully proved. The work of draining many roads is limited because of the difficulty of securing proper outlets for the drains. Nothing will obviate this except hard and sometimes expensive work. It can never be expected that such roads will be good until they are drained in such a way that storm-water will pass quickly from the surface, and the sub-stratum be kept firm by thorough under-drainage.

www.ingramcontent.com/pod-product-compliance
Lightning Source LLC
Chambersburg PA
CBHW032159010726
47493CB00008BA/2748